Water-Quality Trading

A Guide for the Wastewater Community

Cy Jones
Lisa Bacon
Mark S. Kieser
David Sheridan

McGraw-Hill

New York Chicago San Francisco Lisbon London Madrid
Mexico City Milan New Delhi San Juan Seoul
Singapore Sydney Toronto

The **McGraw·Hill** *Companies*

Cataloging-in-Publication Data is on file with the Library of Congress

Copyright © 2006 by the Water Environment Federation. All rights reserved.
Printed in the United States of America. Except as permitted under the United
States Copyright Act of 1976, no part of this publication may be reproduced or
distributed in any form or by any means, or stored in a data base or retrieval
system, without the prior written permission of the publisher.

Water Environment Research, WEF, and *WEFTEC* are registered trademarks of
the Water Environment Federation.

1 2 3 4 5 6 7 8 9 0 DOC/DOC 0 1 0 9 8 7 6 5

ISBN 0-07-146418-2

*The sponsoring editor for this book was Larry S. Hager and the production
supervisor was Pamela A. Pelton. It was set in Sabon by Lone Wolf Enterprises,
Ltd. The art director for the cover was Anthony Landi.*

Printed and bound by RR Donnelley.

This book was printed on recycled, acid-free paper containing
a minimum of 50% recycled, de-inked fiber.

McGraw-Hill books are available at special quantity discounts to use as premiums
and sales promotions, or for use in corporate training programs. For more information, please write to the Director of Special Sales, McGraw-Hill Professional, Two
Penn Plaza, New York, NY 10121-2298. Or contact your local bookstore.

Water Environment Federation

Founded in 1928, the Water Environment Federation (WEF) is a not-for-profit technical and educational organization with members from varied disciplines who work toward the WEF vision of preservation and enhancement of the global water environment. The WEF network includes water quality professionals from 76 Member Associations in 30 countries. For information on membership, publications, and conferences, contact

Water Environment Federation
601 Wythe Street
Alexandria, VA 22314-1994
(703) 684-2400
www.wef.org

Water Environment Research Foundation

The Water Environment Research Foundation (WERF) is a nonprofit organization that helps utilities and corporations preserve the water environment and protect public health by providing science and technology for enhanced management of our water resources. WERF subscribers are utilities and municipalities, environmental engineering and consulting firms, government agencies, equipment manufacturers, and industrial organizations, all with a common interest in promoting research and development in water quality science and technology.

Water Environment Research Foundation
635 Slaters Lane, Suite 300
Alexandria, VA 22314-1177
Tel: (703) 684-2470
Fax: (703) 299-0742
www.werf.org
werf@werf.org

Contents

Chapter One: An Introduction to Water-Quality Trading—*Cy Jones*

Chapter Two: General Conceptual Models for Water-Quality Trading—*Cy Jones*

Chapter Three: Water Quality and Wastewater Treatment Plants—*Cy Jones*

Chapter Four: An Economic Framework for Evaluating Trading Opportunities—*Lisa Bacon*

Chapter Five: Trade—*Cy Jones*

Chapter Six: Science, Data, and Analytical Needs—*David Sheridan*

Chapter Seven: Societal Requirements for Water-Quality Trading—*Cy Jones*

Chapter Eight: Gaining Public Acceptance— *Mark S. Kieser*

List of Figures

List of Tables

Foreword

For much of the 20th century, the waters of the United States were in crisis—the Potomac River was an embarrassment to the country and its capital, Lake Erie was dying, and the Cuyahoga River had burst into flames. Many of the nation's rivers and beaches seemed to be little more than open sewers. In 1972, the Clean Water Act (CWA) was passed to restore and maintain the integrity of the nation's waters. This historic legislation launched an all-out assault on water pollution. It called for reductions in pollution discharges and established interim goals for protecting fish and improving wildlife and recreational uses and ambitious ultimate goals such as attaining "zero discharge" of pollutants.

Over the past 30 years, the primary focus of the CWA has been on the control of pollution from point sources. The application of technology-based discharge requirements and water-quality based effluent limits through the National Pollutant Discharge Elimination System (NPDES) permit program has achieved tremendous success and remains critical to maintaining water-quality goals. However, despite these accomplishments, approximately 40% of rivers, 45% of streams, and 50% of lakes that have been assessed still do not support the beneficial uses, such as swimming, for which they had been designated.

Today we face an array of challenges somewhat different than those we faced in 1972. Nonpoint-source pollution, not directly regulated by the CWA, is now the most significant source of water pollution. Nutrient and sediment loads from agriculture, urban runoff, atmospheric deposition, and wastewater treatment plants are significant contributors to such large-scale, water-quality problems as the "dead zone" in the Gulf of Mexico and the diminished biology of the Chesapeake Bay. Finding solutions to these complex water-quality problems and the smaller-scale, but equally important, local ones will require greater efficiency and innovative approaches. Meeting water-quality standards, while accommodating growth and development, will be a constant challenge. At the same time, any new approaches we undertake must remain aligned with the core CWA programs that form the backbone of the nation's efforts to control water pollution.

Taking a market-based approach to water-quality goals is one such innovation that potentially offers greater efficiency. Water-quality (or watershed-based) trading would allow a discharger to meet its regulatory obligations by using pollutant reductions created by another discharger with lower pollution-control costs. Taking advantage of such cost differentials and capitalizing on economies of scale can reduce the overall costs of controlling pollution. As part of President Clinton's Clean Water Initiative, a proposal to reauthorize the Clean Water Act in 1994, the U.S. Environmental Protection Agency (U.S. EPA) estimated that the potential cost savings associated with water-quality trading ran from a low of $658 million to a high of several billion dollars (U.S. EPA, 1994). Nitrogen trading among point sources in Connecticut was predicted to save over $200 million during a 14-year period of controlling discharges to Long Island Sound. After its first year, the program has achieved over $1 million of surplus nitrogen credits and cut nearly six years off the projected timeline for meeting water-quality standards (Johnson, 2003).

In the past, one of the greatest impediments to trading has been the lack of information and the relatively small number of actual trades, which has helped create the skeptical attitude that, "If trading is so great, why aren't there any trades?" However, progress has slowly but steadily been made to overcome this lack of information and skepticism, beginning with U.S. EPA's issuance of Effluent Trading in Watersheds: Policy Statement in 1996 (http://www.epa.gov/owow/watershed/trading/tradetbl.htm) and *Draft Framework for Watershed-Based Trading* (EPA-800/R-96-001), also in 1996 (http://www.epa.gov/owow/watershed/trading/framwork.html). Over the past decade, a number of studies and pilot programs have been completed that demonstrate the environmental and economic benefits that trading offers.

Given the potential benefits of trading, one could ask, "Why isn't this thoroughbred racing along?" But perhaps the horse has come out of the gate in fine fashion and is just unfamiliar with the course. The thoroughbred need only get its bearings.

Water-quality trading is ready to move into large-scale implementation. The financial incentives are clear, and the policy groundwork has been laid. On January 13, 2003, U.S. EPA issued its final Water Quality Trading Policy (http://www.epa.gov/owow/watershed/trading/finalpolicy2003.html).

While this policy is built on the CWA and its implementing regulations, it signals U.S. EPA's strengthened and broadened support for water-quality trading. Existing regulations provide the legal framework for incorporating trading into water-quality standards, water-quality management plans, NPDES permits issued to point sources, and total maximum daily loads established to restore impaired waters. The policy provides guidance to states, tribes, and sources on how trading can be aligned with and incorporated to these programs. The challenge will be to overcome the many implementation issues that are sure to arise.

U.S. EPA's issuance of a new water-quality trading policy marked a milestone along the road to cleaner water. It signaled the agency's commitment to trading and other market-based initiatives as innovative tools to help achieve the goals of the CWA and encouraged states and tribes and the water-quality community, as a whole, to develop and implement such approaches.

G. Tracy Mehan, Assistant Administrator for the Office of Water until December 2003, recently said, "Water-quality trading is an idea whose time has come." Now it is up to us to make it a reality and to achieve the environmental and economic benefits it offers.

Dave Batchelor
Senior Policy Advisor on Water-Quality Trading
U.S. EPA Office of Water, October 2001 to October 2003

References

Johnson, G. (2003) Reducing Hypoxzia in Long Island Sound: The Connecticut Nitrogen Exchange. Presented at the National Forum on Water Quality Trading, Chicago, Illinois, July.

U.S. Environmental Protection Agency (1994) *President Clinton's Clean Water Initiative: Analysis of Benefits and Costs*; EPA-800/R-94-002; Washington, D.C.

Preface

Over the past decade, the Water Environment Research Foundation (WERF) sponsored five water-quality trading research projects across the country—the Connecticut Long Island Sound nitrogen trading program; the Cherry Creek, Colorado, phosphorus trading program; the Kalamazoo, Michigan, phosphorus trading program; an assessment of the potential for nitrogen trading in Maryland's portion of the Chesapeake Bay watershed; and the Fox/Wolf Basin phosphorus trading program in Wisconsin.

This book was inspired by these projects as well as efforts by many individuals and groups across the country to develop water-quality trading programs. The Water Environment Research Foundation provided funding to support the development of this book as well as oversight and review of the final product. It is a joint effort of WERF and the Water Environment Federation (WEF), with the goals of sharing information and lessons learned with the water-quality community, furthering the development of trading programs, and contributing to the achievement of the nation's water-quality goals.

A work such as this one is invariably the result of the dedication, hard work, and contributions of a great many people. This is truly the case with this book and I am deeply grateful to them all, whether I remember to mention them here or not. I would first like to acknowledge and thank my contributing authors, Lisa Bacon, Dave Sheridan, and Mark Kieser, who were a pleasure to work with and who, in addition to contributing chapters, did much to improve the book in general.

I am deeply grateful to Margaret Stewart, Linda Blankenship, and WERF for giving me the opportunity to undertake this effort and for providing unfailing support and assistance through the many trials and tribulations of producing the manuscript. Thanks are also due to Lorna Ernst at WEF for her adroit handling of the book project. Reaching far back into the past, I would also like to thank Raynetta Grant for getting me entangled in WERF's water-quality trading research efforts nearly a decade ago.

I cannot say enough about the book's peer reviewers—Lynda Hall, Paul Stacey, Leon Holt, Norm LeBlanc, Rob Greenwood, Allison Wiedeman, Julie Vlier, and Jim Keating. Their efforts at identifying the many flaws and omissions in the early drafts and in making numerous beneficial suggestions went far beyond the call of duty. They proved to be an extremely knowledgeable, hard-working, and generous group and the vast amount of insight and information they provided improved and enriched the book in countless ways. I am forever in their debt. Many other people provided information, inspiration, or support as well, chief among them Dave Batchelor, Mahesh Podar, Paul Calamita, Gary Johnson, Bob Moore, Rhonda Sandquist, and Andy Fang.

I cannot conclude without thanking my co-workers at the Washington Suburban Sanitary Commission, whose support has been invaluable to me. Finally, and most importantly, thank you Carol and Matt. I couldn't have done it without you.

Cy Jones

About the Authors

Cy Jones

Cy Jones is the Regulatory Planning and Compliance Manager for the Washington Suburban Sanitary Commission in Laurel, Maryland, where his responsibilities include regulatory compliance, permitting, enforcement issues, and policy development on a broad range of environmental and regulatory issues. He has a B.S. in zoology and an M.S. in environmental engineering from the University of Iowa. He served as Chair of the Water Environment Research Foundation's Project Subcommittee for five water-quality-trading demonstration projects funded by the Water Environment Research Foundation. As Chair of the Nutrient Committee of the Maryland Association of Municipal Wastewater Agencies, he represented Maryland point sources on the Chesapeake Bay Program Nutrient Trading Task Force that developed the Nutrient Trading Fundamental Principles and Guidelines published by the U.S. Environmental Protection Agency and is currently assisting the Maryland Department of the Environment in determining the best role and structure for a nutrient trading program to help achieve the state's Chesapeake Bay Tributary Strategies. He is a Past President of the Chesapeake Water Environment Association and past member of the Water Environment Federation's Board of Directors.

xxiv ■ *Water-Quality Trading: A Guide for the Wastewater Community*

Lisa Bacon

Lisa Bacon is a Principal Technologist with CH2M Hill in Herndon, Virginia. She has 16 years of experience providing strategic and technical support to utilities, watershed communities, and state and federal agencies, helping them identify and implement ways to increase the cost-effectiveness of watershed management programs. She wrote several early U.S. Environmental Protection Agency (U.S. EPA) trading studies and was the lead contract author for U.S. EPA's 1996 *Draft Framework for Watershed-Based Trading*. She was the Principal Investigator for the Water Environment Research Foundation's (WERF's) "Nitrogen Credit Trading in Maryland: A Market Analysis for Establishing a Statewide Framework" project and is currently the Principal Investigator for the WERF project featured in the workshop, "Water Quality Credit Trading: Tools for Assessment and Implementation." Lisa has supported trading pilot studies for watersheds in Alabama, Colorado, Florida, Michigan, Mississippi, Pennsylvania, and Washington, and is currently supporting trading and watershed permitting initiatives in California, Maryland, Oregon, Texas, and Virginia.

Mark S. Kieser

Mark S. Kieser is principal of Kieser & Associates, an environmental science and engineering firm in Kalamazoo, Michigan, that specializes in water resources research, watershed management, water-quality modeling, and new program development. Mr. Kieser has a bachelor's degree in biological sciences from Wittenberg University in Springfield, Ohio, and master's degree in biological sciences from Michigan Technological University in Houghton, Michigan. Mr. Kieser led one of the five U.S. EPA supported water-quality trading projects in the United States in the late 1990s. He served on the State of Michigan Water Quality Trading Workgroup that developed the framework for Michigan's 2002 water-quality trading program. Mr. Kieser has also served as Acting Chair of the Environmental Trading Network since 2001, an internationally recognized clearinghouse for market-based environmental programs. He is currently directing other regionally and nationally recognized projects in watershed management and urban stormwater research.

David Sheridan

David Sheridan is a principal in Aqua Cura, a consulting engineering firm in Camp Hill, Pennsylvania, specializing in water management. He is trained in civil engineering, with a B.S. from the University of Pittsburgh, Pennsylvania, and M.S. and Ph.D. from Penn State University, University Park, Pennsylvania, and is a registered professional engineer in ten states and the District of Columbia.

About the Reviewers

Rob Greenwood

Rob Greenwood, Vice President and Partner at Ross & Associates Environmental Consulting, Ltd., has 21 years of experience designing and managing complex environmental and public health programmatic and policy projects. He has substantial responsibility for conducting analysis and facilitating collaborative stakeholder processes geared to consensus building for durable program and policy change. Since 1995, Mr. Greenwood has used his business finance background to develop a highly successful "market-based instruments" practice at Ross & Associates. This practice focuses on assisting public sector environmental management agencies in using such tools as water-quality trading, voluntary environmental improvement initiatives, and flexible permitting to lower environmental protection costs and encourage "beyond compliance" environmental performance. He has direct experience supporting water-quality trading market analysis, design, development, and/or implementation for several watersheds, including the Puyallup–White River in Washington and the Lower Boise, Middle Snake, Snake River–Hells Canyon, and Portneuf Rivers in Idaho. Mr. Greenwood is the primary author of U.S. EPA Region 10's "Water Quality Trading Assessment Handbook" and U.S. EPA "National Water Quality Trading Assessment Handbook."

Lynda Hall

Lynda Hall is a senior policy analyst with U.S. EPA Office of Wetlands, Oceans, and Watersheds, where she is U.S. EPA's subject matter expert in water-quality trading. Ms. Hall is the coauthor of U.S. EPA's 2003 Water Quality Trading Policy and Chair of U.S. EPA's internal workgroup, the Water Quality Trading Network.

Leon Holt

Leon Holt is the Utilities Pretreatment Manager for the Town of Cary, North Carolina. He is active with the Lower Neuse Basin Association and the newly formed Neuse River Compliance Association. Mr. Holt developed and coordinated a research proposal with North Carolina State University to the Water Environment Research Foundation and U.S. EPA for better understanding fats, oils, and grease (FOG) induced sanitary sewer overflows, the ultimate goal of which is to establish technically based local limits rationale for discharges from food service facilities and as a supplement to nutrient and pathogen total maximum daily load modeling. Mr. Holt is an instructor for the Water Environment Federation for its FOG workshops held around the country. He attended Barton College in Wilson, North Carolina, and graduated with a B.S. in biology. He later attended Fayetteville Technical Institute and received an associate of applied science degree in environmental engineering technology.

Norman E. LeBlanc

Norman E. LeBlanc has more than 30 years experience in the field of water-quality management. As Chief of Technical Services for the Hampton Roads Sanitation District, Virginia Beach, Virginia, his primary responsibilities include National Pollutant Discharge Elimination System, biosolids- and air-permitting activities for nine major and four minor publicly owned treatment works. He has been an active participant in the Chesapeake Bay Program in the development and implementation of water-quality criteria for the control of nutrient and suspended solids effects on living resources of the Bay. He serves as Chair of the Water Quality Committee for the Association of Metropolitan Sewerage Agencies and for the Virginia Association of Municipal Wastewater Agencies and has served on numerous other committees and boards. His interests include total maximum daily loads, nutrients, whole effluent toxicity, and toxicity of metals to aquatic organisms. His education is in physical oceanography.

Paul E. Stacey

Paul E. Stacey is a supervising environmental analyst with the Connecticut Department of Environmental Protection's (CTDEP's) Bureau of Water Management, Planning, and Standards Division. Mr. Stacey has served as state coordinator for the Long Island Sound Study since he was hired in 1985 and also supervises CTDEP's nonpoint-source pollution control program. Mr. Stacey was previously employed by the Academy of Natural Sciences in Philadelphia, Pennsylvania, for eight years.

Julie Vlier

Julie Vlier, P.E., is Manager of Water Quality Services at URS Corporation. She has more than 20 years of experience in water resources engineering, water-quality engineering, and management. Ms. Vlier has worked with numerous clients in the areas of water resources planning, water supply development, water reuse, utility planning, source water protection, total maximum daily load development and implementation, watershed-based trading, watershed-management planning, and water-quality monitoring. Ms. Vlier has provided clients with National Pollutant Discharge Elimination System permitting assistance, including the evaluation of regulatory requirements and development and negotiation of permit conditions with various state agencies. Ms. Vlier has also led watershed-based-trading projects in U.S. EPA Region 8, specifically for phosphorus trading in the Cherry Creek watershed (Colorado), and selenium trading in the Colorado River. On behalf of her clients, she has also provided project management and technical oversight in the areas of wastewater utility planning, Clean Water Act compliance, site application approval, and wastewater design. Ms. Vlier recently completed her second term on the Colorado Water Quality Control Commission, serving as Chair of the nine-member board that promulgates water-quality standards and classifications to protect beneficial uses of state waters.

Allison Wiedeman

Allison Wiedeman received her environmental engineering degree from Vanderbilt University in Nashville, Tennessee, in 1980 and has been working since then with the U.S. EPA. At their headquarters office, she was project manager for the development of national water pollution regulations for energy-related industries. She worked at the U.S. EPA Chesapeake Bay Program for the past 9 years, where she was in charge of directing activities to accelerate the restoration of the Bay through technological innovation and implementation. Her efforts have included development of an accurate tracking program for point-source discharges throughout the Bay watershed; development of nutrient-trading guidelines for the Bay watershed; working with municipalities, industries, and state and local governments to develop programs to reduce their point-source pollutant loading; and an extensive study to determine the costs and economic effects of point- and nonpoint-source nutrient-reduction efforts for the Bay watershed as a whole. She has just returned to U.S. EPA headquarters as the Chief of the Rural Branch in the Office of Water's Water Permits Division. There, she is in charge of the national implementation of the new regulations for concentrated animal feeding operations and policy on permitting of other rural discharges.

CHAPTER ONE

An Introduction to Water-Quality Trading

Introduction

"Why isn't this thoroughbred racing along?" is the insightful question posed in the foreword to this book. There has been a mixed reaction to water-quality trading by some in the wastewater and regulatory communities. Many are skeptical of the alleged benefits, some are leery of unseen pitfalls, and others regard trading as a dangerous erosion of regulatory authority and the ability to control water pollution. In truth, the "thoroughbred" is not "racing along" yet for a number of reasons, one of which is that it is not yet clear enough for some that it is indeed a thoroughbred, and not a swayback nag, or worse, a Trojan horse soon to disgorge vast new regulatory requirements for point-source responsibility for nonpoint-source controls, or hordes of free-wheeling traders who would sack and burn the nation's core water-quality programs.

This book is written from the perspective of the wastewater community; its authors and contributors are largely from this commu-collective feeling of optimism among the authors about the benefits of water-quality trading and its future role in water-quality management. The book was written to help overcome the information gap and the skepticism described in the foreword and hopefully to help guide the wastewater community to the realization of the vision presented there. The authors believe trading is, as it has been proclaimed, a thoroughbred.

Because of this wastewater community perspective, the book will also occasionally touch on the opinions and reactions of the regulated community to the water-quality management programs that affect it. This is especially true of Chapter 3, which presents a comprehensive overview and discussion of water-quality management regulatory programs. The discussions of the perceived strengths and weaknesses of these programs are included to help wastewater treatment plants (WWTPs) more fully understand the artfulness and uncertain science of water-quality management and their opportunities for influencing and improving both the science and the regulatory process. (Note: The term "wastewater treatment plant" [WWTP] applies to both public and private plants treating domestic wastewater and industrial wastewater treatment plants. The term "publicly owned treatment works" [POTW] is also used in this book. POTW is the Clean Water Act [CWA] term for publicly owned municipal WWTPs. The term WWTP is generally used in the book because it is a more encompassing term; water-quality trading is not restricted to POTWs.) If these discussions are perceived as criticisms, it is hoped that the perception is of constructive criticism, for that is truly the spirit that is intended. The authors have immense respect for the U.S. Environmental Protection Agency (U.S. EPA) and the state regulatory agencies and fully appreciate the enormity and difficulty of the tasks they face in protecting the aquatic environment.

As members of the wastewater community, the authors also welcome this opportunity to commend U.S. EPA for its leadership and vision in initiating and furthering the development of water-quality trading. The departure from strict command-and-control approaches and the willingness to display flexibility in solving water-pollution problems has done much to facilitate the spirit of cooperation and partnership that is prevalent today wherever trading programs are being developed. Over the past decade, many people at the federal, state, and local levels have contributed to this spirit.

What is Water-Quality Trading?

The conventional approach for controlling discharges of a given pollutant from WWTPs has been that every plant in a watershed must meet its allocation, or mass loading limit, using its own treatment process (and upgrading if necessary). Water-quality trading is a strategy that hopefully achieves the same water-quality results, while avoiding the universal requirements of the conventional approach. What is water-quality trading? Unfortunately, there is no simple answer to the question. There is no consensus opinion about what activities constitute trading and what activities might more properly be called something else.

It would be useful to start by presenting an example from the Clean Air Act (CAA). The sulfur dioxide (SO2) trading program has become famous for its success in reducing acid-rain-causing emissions at much lower costs than predicted. The program is described in the next section, which is based on a summary of the SO_2 allowance market presented by Shabman et al. (2002).

The Sulfur Dioxide Allowance Market

In 1990, CAA amendments that were designed to reduce emissions causing acid rain, notably SO_2, were adopted. The amendments covered emissions by large utilities and industrial boilers, and emission reduction requirements were imposed in two phases. Phase I began in 1995, covered 263 sources, and required a total reduction of 3.2 million metric tons (3.5 million short tons) per year from 1980 levels. Phase II began in 2000 and brought the total number of regulated sources to 2000. A cap of 8.2 million metric tons (9.0 million short tons) per year was set for all sources, and U.S. EPA allocated the cap using a formula set by the CAA. An emission allowance was defined as one ton per year and most of the allowances went to existing sources based on prior emissions. Sources were free to use, sell, or indefinitely bank their allowances. All other requirements on emissions (emission rates, technology-based controls, etc.) were eliminated. Sources were required to install continuous SO_2 monitors and report emissions to U.S. EPA. A source can meet its allocation by reducing its emissions, using allowances it has previously banked, or by buying allowances.

Substantial cost savings have resulted compared to what would have been incurred under the previous regulatory approach (Ellerman et al., 2000; cited in Shabman et al., 2002). Ellerman and others estimated that, in Phase I, the sources saved $358 million per year from 1995 to 1999, and that Phase II savings for the years 2000 to 2007 will be $2.3 billion per year. This produces a total savings of $20.2 billion over this 13-year period. In addition, the figure does not include U.S. EPA's savings because of its decreased regulatory burden. Removal costs, if each source were to reduce its emissions, were originally estimated at $827 to $1653 per metric ton ($750 to $1500 per short ton). The price of one SO_2 credit (one short ton) now fluctuates between $100 and $200. Much of the success of this program can be attributed to the economic incentives for the sources to improve management and technologies.

Because of the differences between the air and water environments, the structure and methods of the SO_2 trading program are not directly applicable to water-quality trading. A simple example can be used to illustrate water-quality trading in its most basic form. Assume that a watershed has two WWTPs (and no significant non-point-source loads). A total maximum daily load (TMDL) sets an annual phosphorus cap for the watershed at 3629 kg (8000 lb) per year. The permitting authority then sets National Pollutant Discharge Elimination System (NPDES) permit limits for the two plants, equally dividing the 3629 kg (8000 lb) of allowable phosphorus load between them. Plant A then determines that it can reduce its annual phosphorus load to 1361 kg (3000 lb) at some moderate capital and operating costs. Plant B, meanwhile, determines that, with operational changes, it could reduce its load to 2268 kg (5000 lb) but would have to build final effluent filters to go below that level. Plant B then determines that it would be more cost-effective to purchase 454 kg (1000 lb) of phosphorus credits from Plant A, at a price arrived at through negotiations, than to build the filters. (Note: Water-quality trading does not actually require any exchange of money or other considerations. Technically, it would be most accurate to refer to trading partners as users and suppliers of credits. This book uses the terms users and suppliers and buyers and sellers interchangeably.) The regulatory agency approves the trade, and the permit limits are adjusted for both plants.

In essence, what has been exchanged or traded is permission from the regulatory agency to discharge 454 kg (1000 lb) of phosphorus per year. These kilograms or pounds of phosphorus have been termed a *discharge allowance*, and *water-quality trading* has been defined as the exchange of discharge allowances (Stephenson et al., 1999).

Note that this example describes one scenario; in reality, there could be many different scenarios under the general definition of trading. The basic principle would still apply; however, each discharging source must meet its baseline requirement (its allocation) either by reducing its pollutant discharge or by acquiring discharge allowance credits. To sell (or supply) credits, a discharger must reduce its own discharge below its baseline allocation by the amount of credits it wishes to sell, so that the total load after the sale remains within the overall allowable watershed load.

Note also that the situation would rarely be this simple in the real world—not all sources of water pollution have allocations (defined discharge allowances) or are regulated, even in impaired watersheds. This has given rise to much discussion about whether some activities labeled as trading (such as the small-scale offset programs described in Chapter 2) should actually be called trading. To some, particularly the economists who have been instrumental in developing and promoting the concept of water-quality trading, the term trade should refer only to those activities where buyers and sellers of water-quality credits seek each other out and conduct their transactions in a marketlike environment (Shabman, 2002). Other arrangements are more properly called offset programs, managed allocations, or something else.

In common usage, the terminology that has developed for trading uses the term in a more encompassing way—any activity that results in shifts in pollution reduction responsibilities or activities is thought of as a method of trading, even if it might be more accurate to call it an offset or something else. This book uses the term water-quality trading in this broad sense and awaits the outcome of the terminology debate for final definitions.

A Brief History of Trading from the Wastewater Perspective

It has been almost a decade since U.S. EPA issued its original trading policy (U.S. EPA, 1996b) and its companion guidance,

Draft Framework for Watershed-Based Trading (U.S. EPA, 1996a). Since then, U.S. EPA has provided financial or staff support for a handful of trading-related studies. However, for a while, there was little strong, proactive support for the advancement of market-based approaches for watershed management. Then, in January 2003, U.S. EPA issued its final Water Quality Trading Policy (U.S. EPA, 2003), revised and updated from two previous versions, and began to express its strong encouragement and support. During this same period, U.S. EPA announced several new watershed-focused efforts, including a requested $21 million congressional appropriation for grants to directly support innovations, including trading and new permitting approaches.

During this same period, support for the development of water-quality trading by the wastewater community was strong, particularly by the Water Environment Federation® (WEF) and the Water Environment Research Foundation (WERF). The Federation and WERF played significant roles in providing forums and resources for research and discussion about water-quality trading. Specialty watershed conferences in 1996, 1998, 2000, 2002, 2003, and 2004 all featured sessions or workshops devoted exclusively to trading. Equally important, WERF, with U.S. EPA support and funding, began five research projects that involved trading program design, market analysis, and program evaluation for trading pilots in Cherry Creek, Colorado; the Fox-Wolf River basin, Wisconsin; Kalamazoo River–Lake Allegan, Michigan; Long Island Sound, Connecticut; and Maryland watersheds tributary to the Chesapeake Bay. Many lessons were learned from these and other early trading programs, and all of these efforts provided key opportunities for the exchange of information and peer support to continue exploring how trading could be used to help improve and maintain water quality.

As noted in the foreword, WWTPs today face a daunting array of challenges: population growth, combined sewer and sanitary sewer overflow control requirements, infrastructure replacement needs, and large funding gaps. Added to this list can be TMDLs. The single biggest driver, to date, for the interest in trading has been the search for more cost-effective ways to achieve TMDL wasteload allocations imposed as regulatory requirements. Sometimes the search is begun by a plant seeking ways to comply with its allocation. Just as often, a watershed group or state agency explores trading as a way to provide incentives for all sources, regulated or not, to help attain water-quality goals. Others see trading as a way to direct resources to activities beyond just reducing pollutant loads, such as stream, habitat, and wetlands restoration.

The CWA provides direct regulation of point sources but only weak and indirect regulation, at best, of nonpoint sources. Because of this, TMDLs also have often become exercises in ratcheting down on point sources, even when they are not a significant cause of water-pollution problems. Wastewater treatment plants frequently end up with as small a wasteload allocation, or load cap, as is politically, financially, or technologically feasible, while load allocations for nonpoint sources remain large, simply because there is no reasonable assurance that significant nonpoint-source reductions could be achieved (for both regulatory and technical reasons). This perception that there are inherent inequities toward point sources in the TMDL program has caused some of the wastewater community's skepticism toward watershed-based trading. It has not helped that some regulators and watershed groups have, at times, described trading as a way to bring point-source dollars to bear on nonpoint-source pollution.

Despite these potential drawbacks, the wastewater community has shown increasing leadership in advancing innovative approaches to watershed management, including trading, credit markets, and creative permit mechanisms.

Interest in trading among the wastewater community is growing; it is increasingly a topic of discussion at state and local wastewater association meetings. Some WWTPs are interested because they know they can generate credits to sell. Others would like to be able to buy credits if it would allow them to defer capital expenditures or lower operating costs. Some WWTPs may not be sure that they want to trade, but recognize the value of having the option available.

In short, interest in trading by the wastewater community is at an all-time high, and WWTPs are willing to contribute to the efforts to develop trading programs.

Purpose and Structure of the Book

As noted earlier in this chapter, this book is written primarily from the point of view of the municipal wastewater community. It attempts to cover the myriad water-quality trading issues in a logically organized and comprehensive manner. While it is intended to be a practical guide for a municipality or WWTP to use in evaluating the potential for water-quality trading and for designing and implementing trades, it is hoped that the approach taken is broad and objective enough to be of value to anyone interested in trading, whether from a wastewater, regulatory, environmental, or public perspective.

While much has been learned over the past decade, largely through trial and error, water-quality trading remains in its infancy. The main reason is that the emerging guiding principle for trading programs is that one size does not fit all. An examination of the various trading programs that have been undertaken or proposed over the past decade reveals that each is unique, with its own incentives, goals, mechanisms, and barriers. A trading solution crafted in one location cannot simply be applied elsewhere without modification. However, while the solutions are unique to each watershed, the issues, incentives, requirements for, and barriers to trading have common themes, and the questions a potential trading project must answer are the same or similar everywhere.

This book presents a framework for addressing these needs. It provides background information, identifies necessary steps, discusses options, identifies potential pitfalls, and suggests the tools that the reader can use in assessing his or her local situation.

There is a perception that water-quality trading is a difficult and uncertain undertaking, fraught with regulatory roadblocks and unseen dangers. A major goal of this book, inspired by the simple, direct, and optimistic language of U.S. EPA's Water Quality Trading Policy (U.S. EPA, 2003), is to help demystify trading, dispel the misconceptions, and to encourage WWTPs interested in trading to move ahead and assess its viability to help realize their water-quality goals. The book is organized as follows:

(1) Chapter 1 introduces the basics of water-quality trading, briefly describes the history of its development, and discusses how to use the book;

(2) Chapter 2 categorizes and describes the basic types of trading programs that have emerged thus far;

(3) Chapter 3 presents the legal and regulatory framework for water-quality management established by the CWA and shows how trading can be fully consistent with all of the requirements of this framework;

(4) Chapter 4 discusses the economics of water-quality trading and presents tools for evaluating the potential economic benefits;

(5) Chapter 5 describes the various elements of a trade;

(6) Chapter 6 discusses data needs, water-quality analysis tools, and the role of scientific uncertainty in water-quality management and trading;

(7) Chapter 7 summarizes trading program elements that would be necessary to satisfy larger societal needs;

(8) Chapter 8 addresses the critical need to obtain public acceptance for trading programs and discusses the means to accomplish this; and

(9) Chapter 9 discusses the final decision on whether to trade.

Taken as a whole, these chapters form a logical, almost stepwise sequence for evaluating and making decisions about trading. In addition, most of the chapters were written in such a manner as to be able to stand alone as useful resources on the subject areas they address. Chapter 3, for example, would be useful for a WWTP superintendent wanting a full understanding of the rationale and strengths and weaknesses of the process that produced the effluent limits in his or her plant's discharge permit, and advice in Chapter 8 on gaining public acceptance is universally applicable to all manner of water-quality management initiatives.

References

Ellerman, A. D.; Joskow, P. L.; Schmalensee, R.; Montero, J.; Bailey, E. M. (2000) Markets for Clean Air: The U.S. Acid Rain Program; Cambridge University Press: Cambridge, United Kingdom.

Shabman, L., Resources for the Future, Washington, D.C. (2002) Personal communication.

Shabman, L.; Stephenson, K.; Shobe, W. (2002) Trading Programs for Environmental Management: Reflections on the Air and Water Experiences. Environ. Practice, 4 (3), 153–162.

Stephenson, K.; Shabman, L.; Geyer, L. L. (1999) Toward an Effective Watershed-Based Trading System: Identifying the Statutory and Regulatory Barriers to Implementation. Environ. Lawyer, 5 (3), 775–815.

U.S. Environmental Protection Agency (1996a) Draft Framework for Watershed-Based Trading, EPA-800/R-96-001. National Service Center for Environmental Publications: Cincinnati, Ohio. http://www.epa.gov/ owow/watershed/trading/framework.html (accessed June 20, 2004).

U.S. Environmental Protection Agency (1996b) Effluent Trading in Watersheds: Policy Statement. http://www.epa.gov/owow/watershed/trading/ tradetbl.html (accessed June 20, 2004).

U.S. Environmental Protection Agency (2003) Water Quality Trading Policy. http://www.epa.gov/owow/watershed/trading/finalpolicy 2003.html (accessed June 20, 2004).

CHAPTER TWO

General Conceptual Models for Water-Quality Trading

Introduction

The trading programs that have emerged thus far can be grouped into three general categories and a fourth, catch-all category. The general categories can be described as managed trading, trading associations, and marketlike trading programs. The catch-all category is small-scale offset programs. It must be stressed that these categories are oversimplifications. The lines between them are sometimes blurry, and hybrid trading programs could easily be designed. In addition, as trading evolves, new types of trading programs may emerge that would not fit easily into any of these categories. However inadequate these categories are, they do serve to help clarify the basic choices available to wastewater treatment plants (WWTPs) and regulatory agencies. Each of them is described in more detail and examples are given in the sections below.

Managed Trading

Managed trading encompasses a broad range of possible programs involving alternative assignments (or acceptance) of pollution reduction responsibilities, actions, and costs.

Managed trading programs have the following general features:

- First, watershed goals are set, and schedules for achieving interim and final goals are determined;

- Analysis of source-control measures for largest reductions, cost-effectiveness, and ability to implement quickly is carried out;

- The best sequence and timing for upgrades is determined;

- Initial upgrades that will produce reductions greater than needed to meet interim goals are chosen;

- If state grant funding is available, award priority is based on optimization results;

- Non-upgraded facilities must purchase credits from upgraded facilities;

■ Additional upgrades are added as needed to comply with goals; and

■ Not all facilities may need to be upgraded.

One of the main features of this type of trading program, as with all trading programs, is that it departs from the conventional approach that every discharger in the watershed is required to reduce its load (or at least the point sources). The conventional approach is replaced with one in which cost-effectiveness and water-quality benefit determine, in some manner, which plants undertake control actions.

What distinguishes the managed trading model from the other models is the central role played by one or more entities independent of the WWTPs. For instance, in the Connecticut Long Island Sound Nitrogen Trading Program described in the following section, both the state and the Nitrogen Credit Advisory Board (NCAB) play centralized roles. While it is true that, under this model, pollution control actions and credit exchanges could be made in a strictly controlled, regulatory manner, the model provides a great deal of room for flexibility, collaboration, voluntary actions, market considerations, and freedom for a WWTP to choose its course. Note that it would also be possible to have this type of managed pollutant reduction without a parallel reallocation of costs, although this would seem to introduce serious equity issues.

The foremost (and, so far, only) example of this type of program is the Connecticut Long Island Sound Nitrogen Trading Program.

Connecticut Long Island Sound Nitrogen Trading Program

Long Island Sound is listed as impaired by both Connecticut and New York. During summer months, the sound suffers from low dissolved oxygen levels in bottom waters at the western end. Through the federal U.S. Environmental Protection Agency (U.S. EPA) Long Island Sound Study National Estuary Program, a comprehensive conservation and management plan was released in 1994. This was followed by a formal total maximum daily load (TMDL) submitted by

Connecticut and New York and approved by U.S. EPA in 2001. The TMDL calls for a 58.5% reduction in total nitrogen loads within the two states from an established baseline. For Connecticut, this meant a reduction of 6.2 million kg (13.7 million lb) of nitrogen per year. The TMDL required a 10% nonpoint-source reduction from urban and agricultural sources in both states. To accommodate this low reduction level from nonpoint sources, Connecticut set its point-source target at approximately a 64% reduction from the baseline. The combined point- and nonpoint-source reductions meet the 58.5% TMDL requirement.

Connecticut's trading program included only the 79 WWTPs located in the state's watershed tributary to the sound. Without trading, each plant would be expected to achieve a 64% reduction in its discharged nitrogen load, regardless of geographic location or relative effect on oxygen levels in the western portion of the sound.

The TMDL established a statewide wasteload allocation to be met by 2014, with 40 and 75% stepdowns in 2004 and 2009, respectively. Connecticut set a schedule more ambitious than the TMDL schedule to assure aggregate compliance; both the statewide allocation and the individual WWTP allocations are stepped down every year. This had the effect of introducing an additional margin of safety in the point-source allocations (Stacey, 2004).

A statewide general National Pollutant Discharge Elimination System (NPDES) permit contains the annual load allocations for each of the WWTPs through 2006, the life of the current permit.

The general permit, issued in late 2001 for 2002 implementation, includes annual permit reductions for the first five-year permit cycle for all 79 WWTPs, and a listing of final wasteload allocation goals for 2014.

A permittee is in compliance if the annual permit limit is met, or if the WWTP has purchased the appropriate number of nitrogen credits to bring the plant into compliance. A permittee has credits to sell if discharge monitoring shows the nitrogen load is below the permit limit. All 79 plants are required to monitor and report to the Connecticut Department of Environmental Protection.

Credits are "equalized" using trading ratios that reflect the relative effect of each discharger, depending on its location in the basin, and distance from western Long Island Sound. All credit sales and purchases are based on these "equivalent credits".

All credits are bought and sold by the state. Pricing and trades are overseen by the NCAB, comprising relevant state agencies and representatives of the regulated municipalities that operate the WWTPs.

Grants of up to 30% for biological nutrient removal (BNR) projects and low-interest loans are awarded to WWTPs from Connecticut's Clean Water Fund (CWF), based on a state priority system that considers nitrogen removal as a scoring criterion.

An optimization analysis was developed to determine the most cost-effective schedule and sequence of BNR upgrades that would meet the stepped-down allocations over the 15-year period. Grants are awarded in accordance with the priorities established by the optimization modeling.

Equivalent credit prices are set by the NCAB based on the capital cost of the BNR portion of projects that have been funded through the CWF, and associated operation and maintenance costs, divided by the kilograms or pounds of nitrogen removed from those projects, and then adjusted by the appropriate trading ratio. An equivalent credit in 2002 was valued at $3.64 per kg ($1.65 per lb).

In July of the year following the trading year (i.e., July 2003 for the 2002 exchange), permittees that did not meet their annual permit limit are required to buy, from the state, equivalent credits necessary to meet the limit.

In August, the state purchases equivalent credits from those WWTPs that did better than their permit limit. In 2002, the aggregate equivalent load limit of 3.02 million kg (6.65 million lb) was bettered by 394 171 kg (869 000 lb), costing the state about $1.4 million to purchase the excess credits. A dry, warm year contributed to the exceptional WWTP performance.

Trading Associations

The creation of trading associations actually predates U.S. EPA's formal attempts to promote water-quality trading. The concept was developed in North Carolina, where the Tar-Pamlico Trading Association was created in 1989 and the Neuse River Compliance Association in 2002, with the NPDES permit effective in January 2003.

Trading association programs are marked by the following features:

- Mass-load limits or goals based on achieving water-quality standards or TMDL wasteload allocations are calculated for the existing point source dischargers in a watershed;

- The dischargers form a trading association and the state agrees to allow the members (or "co-permittees") to aggregate their individual allocations into a single association allocation;

- The association is free to meet the allocation in any manner it sees fit;

- The trading association signs an agreement with the state through which the association becomes the entity responsible for compliance with the allocation;

- Membership in the association is voluntary; any WWTP not joining the association would be responsible for complying with its individual wasteload allocation;

- Membership in a trading association should neither disqualify a WWTP from state cost-share grants, if a cost-share program exists, nor should it make it ineligible for low-interest state revolving fund loans from the state;

- The ability to acquire nonpoint-source credits can be made available to the association if the state has developed a point–nonpoint-source trading program.

The main benefit of the association model lies in aggregating the individual load allocations into a collective total, as opposed to the conventional approach that every WWTP in the watershed must meet its allocation (by upgrading, if necessary). Within the association, plants discharging below what would be their individual allocation would offset plants discharging above theirs. When the association could no longer meet its allocation with its existing facilities, the most cost-effective upgrade that would produce the necessary reductions would be selected by the association members. The manner in which the association allocated costs among its members would be solely up to the association. Presumably, it would develop bylaws and internal operating rules. The Tar-Pamlico Trading Association, however, has operated, to date, under a "gentleman's agreement" that BNR capabilities will be added at member plants whenever normal expansions or upgrades are done.

Another benefit of a trading association is the potential for cooperation and collaboration among association members. It is likely that treatment plant operators and engineers would exchange knowledge and experience and assist each other in optimizing their plants' nutrient removal performances because it would be in their best interests to do so.

The Tar-Pamlico Trading Association and the Neuse River Compliance Association are described in more detail in the following sections.

Tar-Pamlico Trading Association

In the 1980s, the Pamlico River Estuary in North Carolina suffered from low dissolved oxygen, algae blooms, and fish kills because of nutrient over-enrichment. The estuary was declared a "nutrient-sensitive water" in 1989, and the North Carolina Department of Environment, Health, and Natural Resources (now known as the Department of Environment and Natural Resources) developed a program to address the impairments. The initial plan targeted WWTPs in the Tar River basin and proposed technology-based permit limits for total nitrogen of 6 mg/L, although the plants contributed only 13% of the total nitrogen load in the basin. In reaction to what they felt would be excessively high costs from this approach, the WWTPs formed an association and proposed an alternative to the state. After extensive negotiations, the Tar-Pamlico Trading Association and North Carolina Division of Water Quality signed an agreement in 1989 that contained the following provisions:

■ Annual load caps for nitrogen and phosphorus were established for the association for the 1989-to-1995 interim period.

■ The annual load caps would decrease annually.

■ The association would optimize the nutrient removal performance of their existing facilities.

■ The association provided some upfront funding for agricultural best management practices (BMPs) and state staff.

■ The association funded the development of a dynamic water-quality model for the Pamlico estuary.

■ The association, as a whole, would be responsible for compliance with the agreement.

If the association were unable to meet an annual load cap, it could purchase nutrient credits from the state by making offset payments to the agriculture cost-share program or by providing funding for additional state staff. The offset funds would go to an agricultural cost-share program, to be used to fund voluntary nutrient reduction activities by participating farmers.

The first five years of operation of the association are known as phase I. In this initial phase, 14 WWTPs joined the association, the association kept its nutrient loads below the caps every year and achieved an overall 20% reduction in nutrient loads, the dynamic water-quality model was developed, and the association funded approximately $1 million in agricultural BMPs (aided, in large part, by a U.S. EPA grant).

Phase II of the program began in 1996. The water-quality model was used to reassess the nutrient loading goals. As a result, the association agreed to a nitrogen cap that constituted a 30% reduction from 1991 loads and a phosphorus cap set at the 1991 load. Regulatory requirements designed to achieve specified nutrient reduction goals for nonpoint sources in the Tar-Pamlico basin went into effect in 2000 and 2001.

To date, the association has been able to comply with the caps. In phase I, all association members undertook optimization studies for nitrogen and phosphorus removal, and BNR was installed at two of the larger WWTPs as they underwent expansion. These two efforts yielded sufficient reductions for the association to stay within its cap each year, despite increases in wastewater flows. Additional information is available at the North Carolina Division of Water Quality Web site at http://h2o.enr.state.nc.us/nps/tarpam.htm.

Neuse River Compliance Association

The Neuse River Compliance Association differs from the Tar-Pamlico Trading Association in one important way—the members are considered co-permittees for NPDES purposes, and the general permit specifies not only the association allocation, but also the individual WWTP allocations. If the association, as a whole, exceeds its allocation, it would be subject to an enforcement action. In addition, any member exceeding its individual allocation would also be considered to be in noncompliance and subject to an enforcement action.

Members will, of course, exceed their individual allocations; otherwise, there would be no trading or even need for the association. If the association remains within its allocation, however, no members would be considered in noncompliance with the general permit.

It is safe to assume that payments between members will occur. If so, they would be handled in accordance with the internal operating rules of the association.

Marketlike Trading

The theory that marketlike trading programs could bring a variety of benefits to water-quality management is derived from experience with air programs and the success of the sulfur dioxide (SO_2)

emissions trading program. Application of the concept to water quality has been under development since the early 1990s (Stephenson and Shabman, 1996) and was embodied in U.S. EPA's first policy statement on the subject (U.S. EPA, 1996). It is thought that if free markets and entrepreneurial expertise could be brought to bear on water-quality problems, then innovation and greater efficiency would result. As one state administer recently put it, "Trading should allow us to distribute a scarce commodity (money) to protect a scarce resource (assimilative capacity) in the most efficient way possible" (Mabe, 2003).

Marketlike trading programs are marked by the following features:

- Dischargers in a watershed are given mass-load limits or goals based on achieving water-quality standards or TMDL wasteload and load allocations.

- The dischargers could meet their individual limits or goals either by reducing their own discharged loads or by buying credits from other dischargers or sources (point or non-point).

- The buyers and sellers of credits would operate in a market-like environment. They would seek each other out, negotiate terms of the transaction and prices, and hold each other accountable for compliance with the trade agreement or contract.

- The market would operate within general rules established by the state.

- Buyers and sellers would be free to execute whatever sales and arrangements they desire, as long as they comply with the regulatory requirements established by the state for the trading program.

There are many possible variations of marketlike trading. One would be the use of commodity exchanges, as is done for the trading of SO_2 air emissions. Commodity exchanges could be created on a watershed basis and become the vehicle by which credit buyers and sellers find each other. Creation of the exchange itself would not obviate the need for trading rules and appropriate contracts between trading partners.

Another mechanism could be the use of completive bidding (Shabman, 2002). An entity wishing to buy credits could issue a request for bids and award a credit purchase contract to the lowest responsible bidder. Potentially, even a state could use this mechanism to achieve pollutant reductions in a watershed.

To date, no true marketlike trading program has fully emerged. In 2002, the state of Michigan adopted comprehensive trading regulations for use anywhere in the state (Michigan Department of Environmental Quality, 2002). These regulations, the first of their kind, are generally referred to as "the Michigan Rules". They were developed to specifically facilitate marketlike trading (Batchelor, 2003). The rules are summarized in the following section and a detailed synopsis of them is presented in Appendix A.

THE MICHIGAN RULES

The Michigan Rules were adopted in November 2002, after several years of development. The following two sections present a succinct summary of the rules and were taken from the Public Hearing Report to the Joint Committee on Administrative Rules, which was submitted to the legislature following public comment on the proposed rules (Michigan Department of Environmental Quality, 2001).

Principal Reasons for the Proposed Rules

"Water-quality trading is a market-based approach to improving water quality on a watershed basis. The proposed rules establish a statewide program that can reduce the cost of achieving and maintaining water-quality standards and implementing programs required under the federal Clean Water Act (CWA). The proposed rules provide greater operational and regulatory flexibility and establish

economic incentives for pollutant reductions greater than those required by federal and state regulations. Participation is voluntary."

Summary of Contents of the Proposed Rules

Nutrient trading and other types of trades may occur on a watershed basis among and between point and nonpoint sources throughout the state. The rules include specific provisions to be consistent with the federal CWA, state water-quality standards, and permit program requirements.

Sources must make pollutant reductions beyond those required by the most protective requirement to generate a credit. A percentage of all reductions are retired. Each trade will result in a net loading reduction and water-quality benefit. The rules contain prohibitions and restrictions to ensure that the use of credits does not result in adverse localized effects. The rules include a number of design elements to maintain the levels of control and margins of safety that have been achieved in practice and to improve water quality, as follows: trading ratios, actual versus allowed baselines, discount factors, and requirements for directional and contemporaneous nutrient trading.

Each source that generates, uses, or trades credits must submit a notice to the department. The department must approve and register the notices before any trading activity can occur. Operational requirements specified in the notices become enforceable when the department approves and registers a notice. The rules do not require that permits be issued to nonpoint sources that engage in trading or that nonpoint-source requirements be incorporated to point-source permits. Nonpoint-source accountability is provided directly under the rules.

The rules include a dual liability scheme for generators and users of credits. Generators of bad or insufficient credits are subject to treble credit damages. Users of bad or insufficient credits would have the burden to show due diligence. The rules include reconciliation and true-up provisions.

The rules require the department to establish and maintain a trading registry and electronic bulletin board; conduct periodic program evaluations; review watershed management plans for trading; respond to citizen petitions; and perform case-by-case evaluations of proposed trades, alternate quantification protocols, and site-specific discount factors.

Lower Boise River Phosphorus Trading Program

Perhaps the Lower Boise River phosphorus trading program in Idaho and the Kalamazoo River phosphorus trading program in Michigan come closest to the marketlike model. This section presents a description of the Lower Boise trading program.

In Idaho, the Snake River phosphorus TMDL provided the impetus that led to the development of a general statewide program for water-quality trading. The Lower Boise River is identified in the TMDL as a significant source of phosphorus loading to the Snake River, and the development of the trading program initially centered on phosphorus loads in the Lower Boise River basin.

The Idaho Department of Environmental Quality released a draft document entitled Pollution Trading Requirements that spells out the requirements for trading in a concise, straightforward manner (Idaho Department of Environmental Quality, 2002). The document was subsequently retitled Pollutant Trading Guidance and will be posted on the Idaho Department of Environmental Quality Web site as guidance when finalized. The requirements and concepts contained in the draft are as follows:

- Trading is voluntary.
- Trading allows parties to decide how to best reduce pollutant loads.
- Both point and nonpoint sources can create and sell credits.

■ Baselines are established for point sources by a TMDL wasteload allocation.

■ No baselines are established for nonpoint sources, but each sale by a nonpoint source must contain a "water-quality contribution" that ensures a net reduction from the sale and compliance with the TMDL load allocation for nonpoint sources.

Trading is done through private contracts between the buyers and sellers. While some of the content of the contracts is recommended in the state guidance, the contracts are private agreements and are not submitted to the state or made available for public review.

To generate credits, nonpoint sources must use BMPs on the state's list of approved BMPs. Load reductions can be either measured directly or estimated.

Trades will be incorporated to point-source NPDES permits.

CONCLUSIONS

Many benefits of marketlike trading have yet to be realized—few trades have occurred. There have been no trades in the Lower Boise or Kalamazoo. This lack of trades is frequently cited by skeptics of trading as evidence that, while the concept of trading may sound good in theory, it is not workable in the real world. However, there are logical and more benign explanations for the lack of trades. First, when a new and lower wasteload allocation is assigned, the invariable first step by a WWTP is to see how much its discharged loads can be reduced by optimizing the operation of existing facilities or by adding low-cost modifications. Only when the limits of this optimization are reached do WWTPs look to other, more expensive options such as trading or capital upgrades. In the Lower Boise River basin, the major reason that no trades have occurred is that the Snake River TMDL was delayed until late 2003 (Hall, 2004). However, trading is predicted to begin soon (Mabe, 2003).

While all the models for trading programs have great potential, many feel that marketlike trading may have the most potential to bring about new innovation and creativity in finding ways to reduce pollutant loadings. However, much more work is needed to fully develop point-source and nonpoint-source marketlike trading.

Small-Scale Offset Programs

Offset programs are defined here as programs where a discharger is required to take some action in return for increasing its discharged load or for not decreasing it to comply with a new wasteload allocation. There have been a number of small-scale offset programs to date. Because of their variety, they are difficult to categorize. Two examples, the Rahr Malting Company NPDES permit in Minnesota and the Wayland Business Center NPDES permit in Wayland, Massachusetts, are described in the following sections.

Rahr Malting Company

For malting process reasons, the Rahr Malting Company decided to build its own WWTP and stop discharging to the municipal system. However, under a TMDL for five-day carbonaceous biochemical oxygen demand ($CBOD_5$) established for the lower Minnesota River in 1988, no wasteload allocation would be available for this new discharge. Rahr Malting proposed that a combination of effluent limits in its NPDES permit and upstream nonpoint-source reductions be used to offset its load. As a result, an NPDES permit was issued in 1997 that contains the following provisions:

■ Rahr Malting may discharge 68 kg (150 lb) per day of $CBOD_5$ just upstream of the TMDL zone, but the load must be offset. Several sources of credits are available: Rahr accepted a phosphorus limit of 2 mg/L rather than the 3 mg/L that would otherwise have been imposed; Rahr agreed to a year-round $CBOD_5$ limit rather than the seasonal limit that would otherwise have been imposed; and Rahr will fund upstream BMPs.

■ The Minnesota Pollution Control Agency (MPCA) established pollutant equivalency ratios for $CBOD_5$, phosphorus, nitrogen, and sediment (e.g., 1 lb of phosphorus equals 8 lb of $CBOD_5$, and 1 lb nitrogen equals 4 lb of phosphorus).

■ A delivery ratio of 2-to-1 for upstream BMP-generated reductions was developed.

■ The MPCA established a set of rules governing the point–nonpoint trades. Among them are the following: the trades must produce equivalent water-quality effects in the TMDL zone; the nonpoint-source reductions must be in addition to those that would have occurred otherwise (e.g., as a result of regulatory requirements); and there must be accountability for the nonpoint-source reductions.

■ Financing of BMPs will be accomplished by Rahr Malting establishing a dedicated fund with an initial endowment of $200,000, with another $50,000 to be provided during the life of the permit. The fund will be governed by a board of directors comprised of Rahr Malting officials, state officials, and members of the public.

■ All BMPs must be approved by the MPCA.

Rahr Malting executed four separate trades. Two involved converting flood-plain agricultural land to natural vegetation, one involved streambank stabilization coupled with livestock exclusion, and one involved streambank stabilization alone (Fang and Easter, 2003). The $CBOD_5$ and phosphorus load reductions that resulted during the first five years of the permit are shown in Table 2.1.

Fang and Easter (2003) estimated the cost of these reductions during the five-year period, in terms of phosphorus reduction, to be $13.54/kg ($6.14/lb), compared to typical wastewater treatment costs of $8.82 to $39.68/kg ($4.00 to $18.00/lb) (Senjem, 1997; cited in Fang and Easter, 2003). Because Rahr does not own a WWTP, no direct comparison is possible; Senjem used typical phosphorus removal costs for plants in the appropriate size range. In addition, Fang and Easter (2003) estimated that if the structural life of the

TABLE 2.1 Phosphorus load reductions resulting from the Rahr Malting Company trades (Fang and Easter, 2003).

	CBOD$_5$ (kg/d [lb/d])	PHOSPHORUS (kg/d [lb/d])
Permit requirement for credits	68 (150)	8.6 (19)
Actual reduction requirement (2-to-1 trading ratio applied)	136 (300)	17.2 (38)
Average reduction (1997 to 2002)	194 (428)	24.5 (54)

BMPs were 20 years, the Rahr cost would drop to $3.44/kg ($1.56/lb) during that period. Kerr et al. (2000) reported that Rahr Malting Company staff stated that the cost savings to Rahr could average $300,000 per year during a 30-year period. Thus far, the Rahr trades must be deemed very successful.

Wayland Business Center

The information in this section is based on a summary of the Wayland Business Center trade presented by Kerr et al. (2000). Additional information was provided by P. Hogan of the Massachusetts Department of Environmental Protection (Hogan, 2004) and L. Carlsson-Irwin, past Chair, Wayland Wastewater Management District Commission (Carlsson-Irwin, 2004).

Raytheon Corporation operated a testing facility in the town of Wayland, Massachusetts, with a 227 125-L/d (60 000-gpd) WWTP discharging to the Sudbury River. Because of high phosphorus loadings, the river is highly eutrophic during low flows.

After closing the testing facility, Raytheon sold the property to Congress Group Ventures (CGV). The new owners sought to develop a portion of the property as the Wayland Business Center. Congress Group Ventures requested that U.S. EPA Region I, which administers the NPDES permit program in Massachusetts, reissue the NPDES permit to it. Region I and the Massachusetts Department of Environmental Protection initially declined to authorize what they considered to be a new discharge, citing the fact that the Sudbury River failed to meet water-quality standards, although it was on the list of impaired waters only for metals (Environomics, 1999). The permit could be issued, however, if it had an effluent limit for total phosphorus of 0.2 mg/L. To comply with that, a new plant would have to be built at a cost of approximately $1 million.

In 1997, Region I suggested that the permit could be issued if CGV allowed surrounding homes and businesses with failing septic systems to hook up and if the plant would accept a phosphorus limit of 0.5 mg/L. Further, a 3-to-1 trading ratio would be used to quantify the septic or wastewater treatment phosphorus load to the river. Region I and the Massachusetts Department of Environmental Protection believe that these two requirements will produce phosphorus reductions equal to or lower than what the 0.2 mg/L effluent limit would. Congress Group Ventures agreed to these terms, and 75 708 L (20 000 gal) of the 227 125-L/d (60 000-gpd) capacity of the plant was set aside for wastewater flow from the new hookups.

The NPDES permit was issued to CGV in 1998. Shortly afterward, the town of Wayland acquired the plant and its permit from the developer. The town arranged to connect up to 34 properties to the plant, most of them commercial establishments with two government buildings and several residences. The property owners pay the Wayland Wastewater Management District Commission fees for treatment, although not all of the properties have yet been connected to the public sewerage system.

Conclusion

Small-scale offset programs offer great flexibility, and one could probably be devised for many common water-quality management problems.

References

Batchelor, D., U.S. Environmental Protection Agency, Washington, D.C. (2003) Personal communication.

Carlsson-Irwin, L., Wayland Water Management District Commission. Wayland, Massachusetts (2004) Personal communication.

Environomics (1999) A Summary of U.S. Effluent Trading and Offset Projects. Unpublished Report Prepared for U.S. EPA Office of Water: Washington, D.C.

Fang, F.; Easter, K. W. (2003) Pollution Trading to Offset New Pollutant Loadings—A Case Study in the Minnesota River Basin. Paper presented at the American Agricultural Economics Association Annual Meeting, Montreal, Canada, July; http://www.envtn.org/docs/MN_case_Fang.pdf (accessed March 19, 2004).

Hall, L., U.S. Environmental Protection Agency, Washington, D.C. (2004) Personal communication.

Hogan, P., Massachusetts Department of Environmental Protection: Boston, Massachusetts (2004) Personal communication.

Idaho Department of Environmental Quality (2002) Draft Pollution Trading Requirements. http://www.deq.state.id.us/water/wastewater/guidance_pollutant_trading_Nov03.pdf (accessed September 22, 2004).

Kerr, R. L.; Anderson, S. J.; Jacksch, J. (2000) Crosscutting Analysis of Trading Programs—Case Studies in Air, Water and Wetlands Mitigation Trading Systems. Research paper prepared for the National Academy of Public Administration, Washington, D.C.; http://www.napawash.org/pc_economy_environment/epafile06.pdf (accessed April 10, 2004).

Mabe, D. (2003) Reducing Phosphorus Loads in Idaho's Lower Boise River: The Role of Trading from a State Perspective. Presented at the National Forum on Water Quality Trading, Chicago, Illinois, July.

Michigan Department of Environmental Quality (2001) Public Hearing Report to the Joint Committee on Administrative Rules, ORR 1999-036 EQ. Michigan Department of Environmental

Quality, Surface Water Quality Division: Lansing, Michigan; http:// www.westgov.org/wga/initiatives/tmdl/scott/mi_report.pdf (accessed April 16, 2004).

Michigan Department of Environmental Quality (2002) Rule Part 30: Water Quality Trading (effective November 22, 2002). Michigan Department of Environmental Quality: Surface Water Quality Division: Lansing, Michigan; http://www.state.mi.us/orr/emi/ arcrules.asp?type=Numeric&id=1999&subId=1999%2D036+EQ &subCat=Admincode (accessed April 16, 2004).

North Carolina Department of Natural Resources (1994) Non-Point-Source Management Program: Tar-Pamlico Nutrient Strategy. North Carolina Division of Water Quality: Raleigh, North Carolina; http://h2o.enr.state.nc.us/nps/tarpam.htm (accessed April 17, 2004).

Senjem, N. (1997) Pollution Trading for Water Quality Improvement—A Policy Evaluation. Unpublished Report, Minnesota Pollution Control Agency: St. Paul, Minnesota.

Shabman, L. (2002) Presentation at the Maryland Trading Forum, Baltimore, Maryland, April.

Stacey, P., Connecticut Department of Environmental Protection, Hartford, Connecticut (2004) Personal communication.

Stephenson, K.; Shabman, L. (1996) Effluent Allowance Trading: A New Approach to Watershed Management. *Water Sci. Rep.*, **2-3**; Virginia Water Resources Research Center: Blacksburg, Virginia.

U.S. Environmental Protection Agency (1996) Effluent Trading in Watersheds: Policy Statement; http://www.epa.gov/owow/watershed/trading/tradetbl.htm (accessed June 20, 2004).

CHAPTER THREE

Water Quality and Wastewater Treatment Plants

Introduction

The purpose of this chapter is to address the question: Is water-quality trading legal? To do so, it must first address the question: Why do wastewater treatment plants (WWTPs) do what they do? This requires an overview of the Clean Water Act (CWA) requirements as they affect WWTPs and the regulatory requirements that stem from the act.

The CWA's overarching goal is to "restore and maintain the chemical, physical, and biological integrity of the nation's waters" (CWA, 1972; Section 101). The act articulated a national goal "that the discharge of pollutants into the navigable waters be eliminated by 1985", and went on to state "that wherever attainable, an interim goal of water quality which provides for the protection of fish, shellfish, and wildlife and provides for recreation in and on the water be achieved by July 1, 1983" (CWA, 1972; Section 101). This interim goal is commonly referred to as "fishable and swimmable" waters. Moving from the federal, through the state, and to the local levels, these broad water-quality goals are implemented through a series of ever-more-specific programs and requirements. Ultimately, every WWTP in the United States is governed, in great detail, by the many legal and regulatory requirements placed on it as a result of the CWA.

This chapter describes how these water-quality goals and associated water-quality standards serve as both goals and the legal framework to achieve them. It describes the various mechanisms through which specific WWTP requirements are derived from the water-quality goals. The strengths and weaknesses of the process are also discussed, so that a WWTP knows when it is useful and appropriate to become involved in the process.

The chapter then turns to the question: Is water-quality trading legal? It shows how trading can be undertaken in a manner that would contribute to attainment of water-quality goals, comply with legal and regulatory requirements, and be fully compatible with all of the associated technical and legal processes. The conclusion of the chapter is that trading is legal.

Water-Quality Standards

While it is tempting to think of water-quality standards as simple measures of chemical concentrations, it is more accurate to recognize them as multifaceted and elegant constructions that serve many purposes. First, they set forth, in a general way, the various goals for a particular water body in terms of its designated uses (but not necessarily all the uses that may be desired). Numeric and narrative criteria are then developed that, if attained, ensure that the water body would be suitable for the uses. These criteria then form the scientific and legal bases for the planning undertaken for the watershed and the pollution control measures imposed on WWTPs and other dischargers. The U.S. Environmental Protection Agency's (U.S. EPA's) Water Quality Standards Handbook (U.S. EPA, 1994) provides a comprehensive overview of this water-quality management strategy and guidance to states and tribes on all aspects of developing and implementing water-quality standards.

Designated Uses

The designated uses in place today are largely generalized goals, which are organized around four main purposes: protection of human health, protection of recreation, protection of aquatic ecology, and suitability of use by agriculture and industry. A state's set of adopted uses may look like the following:

■ Recreation (primary and secondary contact);

■ Protection and propagation of fish and wildlife;

■ Public water supply; and

■ Agricultural and industrial water supply.

A few states have recently moved to refine their designated uses (see the Ohio example below), adding more specific sub-uses, for example; however, most states have very simple lists.

Water-Quality Criteria

Once a water body is designated for its highest attainable uses, the next step is to develop the numeric and narrative measures that would define attainment. These water-quality criteria are scientifically derived statements of the chemical and physical conditions that are needed to protect each of the adopted uses. Typical examples familiar to WWTPs include dissolved oxygen, pH, temperature, ammonia, and metals. Some states have also adopted numeric criteria for nutrients and turbidity. Numeric criteria have also been adopted for a large number of toxic substances, including many of the priority pollutants.

In some cases, the criteria do not directly address the impairing substance but, instead, are for secondary effects. The prime example of this is criteria for nutrients. Many states do not yet have nutrient criteria, but may have criteria for dissolved oxygen, chlorophyll *a*, or water clarity, and consider nutrients to be the primary cause when they are violated. Nutrient criteria may become more common in the near future, however. U.S. EPA is encouraging and helping states to better quantify protective nutrient levels in their standards. It developed National Strategy for the Development of Regional Nutrient Criteria (U.S. EPA, 1998b), published nutrient criteria for fourteen ecoregions in 2001, and issued guidance for states and tribes to use in the development of water-quality standards for nutrients (U.S. EPA, 2001c). Few states have done so as of this writing.

U.S. EPA is responsible for scientifically developing and publishing numeric criteria recommendations. A separate criterion document containing the scientific rationale for the recommended number is published for each pollutant. U.S. EPA first published all of the criteria documents together in book form in 1986, officially titled *Quality Criteria for Water* (U.S. EPA, 1986), but commonly referred to as the "Gold Book" because of the color of the cover. The book should be used with care, however. A new criterion may have been adopted by U.S. EPA, and a revised criterion document issued since its publication. According to U.S. EPA, the criterion document is the official guidance document, not *Quality Criteria for Water* (U.S. EPA, 1994; p 3-1). Many new criteria have been published since the publication of *Quality Criteria for Water*. The latest set of criteria recommendations can be found at http://epa.gov/ waterscience/standards/wqcriteria.html (accessed July 3, 2004). U.S. EPA also announces updates to criteria through notices in the Federal Register.

When assessing its ability to comply with a water-quality criterion, a WWTP should not rely on *Quality Criteria for Water* (U.S. EPA, 1986) to determine the criteria for its receiving water. The WWTP should obtain information on the criteria for its receiving water directly from its state regulatory or environmental agency.

The numeric chemical criteria are, in many cases, very conservative. They are generally based on the chemical's effects on the most sensitive aquatic species, or conservative bioaccumulation assumptions are made; they are generally under worse-case conditions, and may have safety built-in factors. In this way, U.S. EPA and the states address the scientific uncertainty about the biological effects of the chemical.

While U.S. EPA criteria can be adopted directly by states for individual waters, they can be modified by states for site-specific conditions, though this probably happens less frequently than it should. States are also free to develop numeric criteria using other scientifically defensible methods. Any criterion developed by a state requires review and approval by U.S. EPA, which has broad discretionary authority to promulgate state water-quality standards if it finds a state's standards to be inconsistent with the CWA.

Designated uses are also protected by more generalized nonnumeric criteria that are designed to prohibit nuisance conditions. Maryland's narrative criteria, for example, contain the following (Code of Maryland 26.08.02.02 [Designated Uses, 1995]):
The waters of this State may not be polluted by:

    (1)   Substances attributable to [wastewater], industrial waste, or other waste that will settle to form sludge deposits that:

        (a) Are unsightly, putrescent, or odorous, and create a nuisance, or

        (b) Interfere directly or indirectly with designated uses;

(2)   Any material, including floating debris, oil, grease, scum, sludge, and other floating materials attributable to [wastewater], industrial waste, or other waste in amounts sufficient to:

(a)   Be unsightly,

(b)   Produce taste or odor,

(c)   Change the existing color,

(d)   Change other chemical or physical conditions in the surface waters,

(e)   Create a nuisance, or

(f)   Interfere directly or indirectly with designated uses.

This type of narrative criteria is designed to be very broad, so that it could be brought to bear on almost any type of unanticipated or unquantifiable water-quality problem.

The CWA attempted to ensure that water-quality standards would be kept current with the latest science by requiring that states review them once every three years, an event known as the triennial review. Wastewater treatment plants should be aware of their state's triennial reviews and take advantage of the opportunity to press for improvements to the standards where warranted.

Issues with Water-Quality Standards

While water-quality standards are powerful tools for defining and achieving water-quality goals, their very comprehensiveness and complexity result in several issues or shortcomings in their development and application. Because these issues can cause problems throughout the cascading chain of water-pollution control activities, WWTPs should be aware of them. They are of special concern when dealing with waters that are identified as impaired for their designated uses.

DESIGNATED USES

When the CWA was passed in 1972, it provided the first comprehensive requirement for states to adopt water-quality standards for all of their waters and submit them to U.S. EPA for approval within six months of enactment. There naturally followed a rush by the states to comply with this requirement. As a result, uses were designated for most water bodies without the benefit of much deliberation or scientific analysis. Two shortcomings resulted from this. First, most designated uses adopted during this period are generalized and vague. Most states have only a few overly broad uses, as illustrated by the above list. States have typically sorted hundreds or thousands of distinct water bodies with widely varying characteristics and ecology into a small handful of generalized uses.

It is clear that water-quality planning and management would benefit from improvement to the designated use characterizations. The National Research Council (NRC), in its assessment of the scientific basis of the total maximum daily load (TMDL) program, observed that "Clean Water Act goals (e.g., 'fishable,' 'swimmable') are too broad to be operational as statements of designated use" (NRC, 2001). The NRC report stated that "assigning tiered designated uses is an essential step in setting water quality standards" (NRC, 2001; p 30). A few states have made some progress toward this goal. Ohio, for example, has adopted the following set of three major use designations, each with subdesignations (Water Body Use Designation, 2002):

(1) Aquatic life habitat.

- State resource water.

- Warm water habitat.

- Exceptional warm water habitat.

- Modified warm water habitat.

- Seasonal salmonid habitat.

- Coldwater habitat.

- Limited resource water.

(2) Water supply.

- Public water supply.

- ■ Agricultural water supply.
- ■ Industrial water supply.

(3) Recreation.

- ■ Bathing water.
- ■ Primary contact recreation.
- ■ Secondary contact recreation.

Before setting use designations, Ohio first ensures that adequate monitoring data are available and then performs a use-attainability analysis that examines site-specific conditions. Ohio's approach was cited by NRC in its report on the TMDL program (NRC, 2001). Ohio's regulations and use designations can be found at http://www.epa.state.oh.us/dsw/rules/3745-1.html (accessed March 22, 2004).

The second problem is that, because most of the original designations were made with little or no investigation of the actual water bodies, little information was available on the physical, chemical, or biological characteristics of the waters for the "highest-use" characterization. As a result, many waters in the United States have inappropriate use designations; inappropriate because they are scientifically invalid or because they are unattainable (AMSA, 2002). Examples include full-body contact recreation for streams that are actually too shallow (GAO, 2003) or warm-water fishery for streams in highly urbanized areas with drastically altered physical habitat and flow regimes.

A final issue is the process of reevaluating existing designated uses and making changes where appropriate. U.S. EPA regulations provide for this through a process known as a use-attainability analysis, or UAA (40 CFR 131.10(g) [Water Quality Planning and Management, 2003]). As part of its 2002 study, the U.S. General Accounting Office (GAO) surveyed all 50 states, and all reported that they had designated uses that needed changing (GAO, 2003).

However, the UAA process is a difficult and uncertain one. In response to the GAO survey, the states reported that the main obstacles to doing UAAs were a lack of the resources and monitoring data that would be necessary, resistance from interest groups and other parties, and uncertainty over the possible reaction of the U.S. EPA region to any proposed changes in designated uses. For this reason, few UAAs have been undertaken and many waters remain mischaracterized.

However, because of the increasingly important role of the TMDL program in water-quality management (described later in this chapter), the need for viable and rigorous UAAs is becoming increasingly important. Perhaps UAAs should be done for most, if not all, water bodies scheduled for TMDL development.

Another problem is determining if a designated use is being attained. Ideally, the numeric criteria adopted to support the designated use would be directly related to the in-stream conditions of interest and, hence, would be sufficient to determine attainment. However, this is rarely the case. The attainment determination would be very difficult to make for physically and biologically complex waters. This is particularly true for the "protection of fish and wildlife"-type of designations. There are trade-offs in the choice of criteria, and each choice creates more uncertainty in some area (NRC, 2001). This problem creates both planning and regulatory difficulties and is discussed further in this chapter and in Chapter 6.

Wastewater treatment plants should be familiar with the designated uses of their receiving waters and of other water bodies in the watershed. The establishment of water-quality criteria, the imposition of effluent limits, decision-making on whether a water-body is impaired, and developing TMDLs could all be adversely affected by inappropriate or unattainable use designations. U.S. EPA itself has acknowledged the need to revisit the scientific basis of designated uses (U.S. EPA, 2002b).

NUMERIC WATER-QUALITY CRITERIA

The strength of numeric criteria is that they are very specific water-quality management tools that can be effective when properly applied. This approach, however, also suffers from some drawbacks that can lead to problems for WWTPs and water-quality managers and regulators.

The first drawback is that water-quality criteria require a large amount of data to properly develop, implement, and monitor for attainment. To develop a criterion, information or research is first needed on the known or suspected effects of the substance of concern on aquatic organisms. Data are then needed on water-body chemistry, physical conditions, and biology. Methods and models must be developed to predict the actual in-stream effects of the substance, and these predictions must be tested before the promulgation of the criterion. Data on pollution sources and loadings are then needed to ascertain the water-body response to various pollutant loads. Once a criterion has been adopted for a water body, some level of ongoing monitoring at various locations is necessary to determine compliance with it.

For states, these data demands are resource-intensive, and few, if any, states have been able to devote adequate resources to this process (GAO, 2000). In many cases, states incorporated U.S. EPA's recommended criteria directly into water-quality standards without regard for the possible need for modifications to make them more applicable to the state's waters. Even where this has been done, many states will still apply one generic state criterion to a wide variety of water-bodies, without regard to site-specific or natural conditions. However, there are some relatively simple tools available to tailor generic criteria to local waters, and many states do, in fact, use them. These include chemical and biological translators and water-effects ratios.

The question naturally arises: Do the adopted criteria for a given water body adequately characterize all of the conditions necessary to ensure that the designated use is protected? In almost all cases, the answer is, understandably, no. Our knowledge of aquatic ecology and the biological response to pollution stresses is inadequate. For the past 30 years, the regulatory focus has been on controlling the discharge of specific substances from point sources, and this has largely been successful. However, attention is being increasingly focused on the need to learn more about ecosystem response and to develop valid and usable biocriteria to judge the health of water bodies. While many states have, indeed, adopted them, the NRC recommended that biological criteria that more closely track the ecological conditions of a water body be developed and used in conjunction with physical and chemical criteria (NRC, 2001; p 50).

Another question that arises in the application of numeric criteria is: Should a single exceedance of the criteria be considered a water-quality standards violation? In most cases (other than for toxic substances), the answer should be no. If the answer was yes, many more waters would be defined as impaired, and additional costly pollution control measures would be imposed without any real water-quality benefit. In very few cases does a single criterion exceedance result in an actual use impairment; in addition, many water bodies naturally exceed a criterion periodically, even without a manmade pollutant load. Instead of reliance on a single number (e.g., a single data point from a grab sample), information is needed on the duration of the excursions, and sufficient data are needed to statistically characterize their frequency of occurrence (NRC, 2001; p 45). Because data are generally not available on the duration of excursions, frequency is the most widely used measure to assess compliance.

Neither the CWA nor U.S. EPA regulations address how compliance with a criterion is actually determined. The U.S. EPA has, however, issued guidance as part of the TMDL process, and states have addressed the issue in various ways (e.g., adding a footnote stating that exceedances because of natural conditions or variability are not a violation of the criterion).

For all of the reasons cited above, GAO concluded that "many [U.S.] EPA criteria are not easily comparable with reasonably obtainable monitoring data" (GAO, 2003), and that states generally find it difficult to modify U.S. EPA criteria because of resource limitations, inadequate data, and uncertainty over the U.S. EPA approval process.

NARRATIVE WATER-QUALITY CRITERIA

The nonspecific nature of narrative criteria provide both their power as regulatory tools and their weaknesses. Their general nature allows them to be applied to almost any water-quality problem, and they are very useful when pollutant-specific criteria are impossible or impractical to develop (such as for fat, oil, and grease, or even nutrient criteria, in some cases). This same generality, however, makes them clumsy and imprecise tools; numeric translators are needed to most effectively apply them. Many states are using numeric translators for parameters such as nutrients, chlorophyll *a*, and sediments. This is generally appropriate; however, it creates the potential for sidestepping the CWA requirement that water pollution control mechanisms be derived from designated uses, using the best available science. In adopting designated uses and water-quality criteria, states must follow the established process for promulgating and adopting regulations, including public review. Unfortunately, sometimes when a state starts using a numeric translator for a narrative criterion, it is a number that has not necessarily been subjected to scientific peer review, or even public review, much less the other rigorous tests of the regulatory promulgation process. The danger in this is the potential use of arbitrary or scientifically unsupported translators with resultant over- or under-protective, water-quality protection measures.

U. S. EPA's Twenty Needs Report (U.S. EPA, 2002c) recognized both the usefulness and shortcomings of narrative criteria. It stated

> …a generalized numeric criterion can overshadow local, reach-specific considerations…in these cases, [regulators] prefer to apply their detailed local knowledge with the flexibility afforded by a narrative criterion. The concept of 'translators'—methodologies to guide the calculation of site-specific numeric targets (not criteria) based on a narrative standard—has potential to become a popular substitute for using rigid, pass/fail numbers in numeric criteria (U.S. EPA 2002c; p 25).

This defense of narrative criteria then gets to the real point and states "This concept needs to establish a defensible track record…".

In fairness, it should be acknowledged that, in cases where numeric criteria are overly generalized and the development of site-specific criteria would be particularly onerous, the use of soundly applied narrative criteria could very well be the superior approach.

Antidegradation

U.S. EPA's regulations require each state to adopt and enforce anti-degradation provisions as part of its water-quality standards (40 CFR 131.12 [Water Quality Planning and Management, 2003]). Antidegradation is both a policy and process designed to protect existing levels of water quality. It is also an increasingly prominent component of the water-quality standards program.

U.S. EPA regulations take a three-tier approach toward implementing the antidegradation program.

(1)   Tier I waters are those where one or more water-quality standard is not attained. In these waters, WWTPs or other dischargers could not increase their discharges of the impairing substances, unless in conformance with an approved TMDL. Discharges of other substances could be increased, however. For example, if a water is tier I because of violations of the bacteria standard, a new discharge of copper could be allowed, as long as it did not result in a violation of the water-quality standard for copper.

(2)   Tier II waters have water quality higher than the applicable water-quality standards, and the antidegradation program requires that this higher level of water quality be protected. The regulations state:

Where the quality of the waters exceed levels necessary to support propagation of fish, shellfish, and wildlife and recreation in and on the water, that quality shall be maintained and protected unless the State finds, after full satisfaction of the intergovernmental coordination and public participation provisions of the State's continuing planning process, that allowing lower water quality is necessary to accommodate important economic or social development in the area in which the waters are located. In allowing such degradation or lower water quality, the State shall assure water quality adequate to protect existing uses fully. Further, the State shall assure that there shall be achieved the highest statutory and regulatory requirements for all new and existing point sources and all cost-effective and reasonable best management practices (BMPs) for nonpoint-source control (40 CFR 131.12(a)(2) [Water Quality Planning and Management, 2003]).

In no case can water quality be lowered to violate an applicable water-quality standard. Thus, applying the tier II requirements to the copper example above, a new or expanding WWTP could discharge copper only upon a showing of socioeconomic needs.

> (3) Tier III waters are waters of the highest quality and are also called "outstanding national resource waters". These waters are given the highest level of protection. Generally, that protection is provided through a prohibition against new or expanded discharges of pollutants.

In practice, antidegradation has been a somewhat dormant feature of the CWA. Generally, antidegradation has not been an issue for tier I waters because point-source discharges have been tightly permitted for over 30 years; hence, there is little chance that these discharges could cause violations of existing water-quality standards.

In 1998, U.S. EPA published a Water Quality Standards Advanced Notice of Public Rule Making (ANPRM) (U.S. EPA, 1998a), announcing U.S. EPA's examination of potential revisions to the antidegradation regulations. In its review of the ANPRM, the Association of Metropolitan Sewerage Authorities noted that the ANPRM "contains surprisingly prescient statements about antidegradation's emerging presence" (AMSA, 2002), and that it contained a number of telling statements such as "[antidegradation] is not being used as effectively as it could be" and that it is "significantly underused as a tool to attain and maintain water quality and plan for and channel important and economic and social development that can impact water quality". Soon after the ANPRM, U.S. EPA placed the rulemaking on hold, where it remains as of this writing. In its Legal Perspective on Antidegradation (AMSA, 2002), AMSA also felt compelled to quote the Water Quality Standards Handbook (U.S. EPA, 1994).

Antidegradation is not a 'no growth' rule and was never designed or intended to be such. [States] may decide that some lowering of water quality in "high-quality waters" is necessary to accommodate important economic or social development (U.S. EPA, 1994; p 4-8).

Derivation of Effluent Limits from Water-Quality Standards

The CWA took two different approaches for establishing controls on point-source dischargers. One was the establishment of uniform, technology-based requirements for all dischargers in the United States, both municipal and industrial. In cases where these controls are insufficient to achieve water-quality standards, more stringent effluent limits designed to achieve them are to be imposed. These two types of limits are known as technology-based effluent limits and water-quality-based effluent limits (WQBELs), respectively.

For WWTPs, the technology-based requirements are known as secondary treatment and are defined as the controls necessary to meet monthly average five-day biochemical oxygen demand (BOD_5) and total suspended solids (TSS) concentrations of 30 mg/L (the so-called "30/30 requirement"). Weekly average limits of 45 mg/L were also established for each of these parameters. In addition, minimum requirements for overall removal rates of 85% were established for both.

Revised U.S. EPA regulations for stormwater management are being implemented in two phases. In phase I, "municipal separate storm sewer system" NPDES permits (known as municipal separate storm sewer [MS4] permits) were required for large and medium municipalities with populations of 100 000 or more. Phase II extends the requirement to smaller jurisdictions with populations of at least 50 000 and population densities of 386 persons per square kilometer (1000 persons per square mile) or more. The MS4 permits for both phases require that the permittees develop and implement stormwater management programs (SWMPs) designed to reduce stormwater-related pollutant loadings to surface waters to the "maximum extent practicable", although the regulations do not define what this is. The SWMPs must include BMPs for six "minimum control measures," along with the development of measurable goals to assess the efficacy of the BMPs. Stormwater permits must also comply with any TMDLs adopted for the receiving waters.

In practical terms, the MS4 program, although legally defined as a point-source program, takes a typical nonpoint-source BMP approach to controlling stormwater-related pollutant loadings. As such, it suffers from the same quantification and uncertainty problems that the management of agriculture-related water pollution does. These issues are discussed in detail in Chapter 5.

Direct industrial dischargers to surface waters are regulated differently. For conventional pollutants (biochemical oxygen demand [BOD] and suspended solids), industries were required to achieve best practicable technology, defined as "average of the best existing performance by well-operated plants within each industrial category" (WEF, 1997). This later became best conventional technology. For toxics and nonconventional pollutants, industrial dischargers must meet the most stringent best available technology economically achievable standard (WEF, 1997).

Today, WWTPs have many water-quality-based effluent limits in their NPDES permits. Instantaneous or weekly, monthly, and sometimes annual average limits on nutrients, ammonia, pH, dissolved oxygen, residual chlorine, or metals all constitute WQBELs. An analysis of the effects of a WWTP's discharge on the water quality of the receiving water must be done to establish effluent limits. States use many different methodologies to do this, ranging from simple to complex, depending on the parameter and the water body. Some examples are:

■ Uniform application of environmentally well-established values, such as effluent pH ranges of 6 to 8;

■ Calculation of oxygen-sag curves downstream of the discharge to determine the levels of BOD_5 (and other oxygen-consuming substances) and dissolved oxygen in the effluent that would prevent violation of the in-stream oxygen criterion under critical low-flow conditions;

■ Simple dilution calculations and comparison of the resulting concentrations at the edge of a mixing zone to water-quality standards;

■ Steady-state, water-quality models;

■ Dynamic water-quality models; and

■ Toxicity testing (this may be a particularly advantageous approach because it could avoid some of the issues of overly-conservative assumptions and additive or multiplicative effects).

Whatever analytical method is used, states must document the predicted water-quality effect of the discharge and present the water-quality rationale for the proposed effluent limit in a fact sheet accompanying the draft permit. In establishing a particular effluent limit, the permitting agency must show that, without such a limit, the discharge would have a reasonable potential to result in violation of a water-quality standard. Hence, effluent limits are derived directly from the water-quality standards of the receiving waters.

This process creates fertile ground for regulators and dischargers to find themselves in disagreement. Wastewater treatment plants should keep the following points in mind:

■ While there must be some scientific basis for a proposed effluent limit, regulatory agencies have a great deal of legal and scientific leeway in establishing it.

■ Scientific uncertainty will always be present in water-quality management (NRC, 2001), so regulatory agencies will be given the benefit of the doubt by the courts.

■ Effluent limits are designed to be protective during rare, worst-case conditions such as extreme low-flow conditions. As such, they will be established in a conservative manner. Wastewater treatment plants, however, can challenge proposed limits when an excess of conservative assumptions in the analysis have a multiplying effect that produces theoretical conditions that would never actually occur.

■ Limits do not have to be fixed for all conditions but can vary with season, flow, or other factors. Ammonia is a good example; its toxicity varies with temperature and pH, and is more toxic to salmonids than warm-water fish species. Hence, many WWTPs have a limit on ammonia discharge during summer months, but a higher one or no limit at all during winter months.

■ Opportunities for public review and comment are available at virtually every step of the regulatory process, and WWTPs should take advantage of this and offer comments on these issues to the promulgating agencies at the appropriate times.

■ Dischargers do have opportunities to mount scientific and legal challenges.

Antibacksliding

The CWA attempts to ensure that effluent limits, once issued in a permit, will not be relaxed in subsequent reissuance of the permit. Section 402 (o) of the Act terms this backsliding and contains provisions to prevent it. It states that

> ...a permit may not be renewed, reissued, or modified...to contain effluent limitations which are less stringent than the comparable effluent limitations in the previous permit (CWA, 1972).

The apparent absolute nature of this prohibition is greatly tempered, however, by the inclusion in 402(o) of a broad range of exceptions, including the availability of information that was not available at the time the original permit was issued. Antibacksliding has generally not been an impediment to changing effluent limits in light of better data or modeling.

One of the exceptions in Section 402(o) pertains to TMDLs.

...a permit may not be renewed, reissued, or modified to contain effluent limitations which are less stringent than the comparable effluent limitations in the previous permit, except in compliance with section [303(d)] (CWA, 1972).

In other words, if water-quality standards were not being attained in the receiving water, effluent limits could be relaxed if the new limits would be in compliance with a TMDL wasteload allocation. If water-quality standards were being attained in the receiving water, then permit limits could be relaxed, as long as consistency with the state's antidegradation procedures were maintained (Batchelor, 2004).

The above discussion notwithstanding, it should also be noted that the antibacksliding rule is complex and poorly understood, and its implications have not yet been fully explored (Calamita, 2004).

Clean Water Act Section 303(d)— Total Maximum Daily Loads

While the initial focus of water-quality management and planning efforts following the adoption of the CWA (CWA, 1972) was on constructing and upgrading publicly owned treatment works to achieve the mandated secondary treatment levels, this short section of the act was added as a backstop to ensure that efforts would not end there. It requires states to

■ Identify those waters not expected to meet water-quality standards, even after the application of technology-based effluent limits, and to establish a priority ranking of those waters;

■ Submit the list of impaired waters to U.S. EPA for review and approval and update the list from time to time;

■ Establish the TMDL of the impairing pollutant necessary to achieve water-quality standards, taking into account seasonable variations and a margin of safety; and

■ Submit such TMDLs to U.S. EPA for review and approval.

The act established a six-month timetable for the submission of the first 303(d) lists of impaired waters by the states, and a general timeframe in which all necessary TMDLs would be established within a relatively few years.

Section 303(d) is seemingly simple. As Houck (1997b; p 10336) noted in his discussion of how and why it came to be included in the CWA, "On its face, there was and is nothing remarkable about § 303(d)". Its simplicity would turn out to be misleading, however.

While long ignored by the states and U.S. EPA, an onslaught of successful litigation beginning in the late 1980s has awakened this sleeping giant (Houck, 1997a). It has now become the main focus of water-quality planning and management in most states as they rush to try to comply with 303(d) requirements and the many consent decrees that have been put into place. As such, 303(d) and the regulations and guidance that U.S. EPA has adopted to implement it will have a major effect on all aspects of water-quality management and planning for the foreseeable future and on the options that WWTPs may have for complying with water-quality requirements.

The remainder of this section describes the TMDL program, from the identification and listing of impaired waters to the implementation of the pollutant allocations identified in TMDLs.

IDENTIFICATION AND LISTING OF IMPAIRED WATERS

The logical framework for water-quality management under the CWA begins with requirements for states to assess the status of their waters on a regular basis and submit the findings to U.S. EPA. The findings are also incorporated into the states' water-quality management plans (WQMPs), "leading to the development of controls and procedures for problems identified..." (40 CFR Part 130.8 [Water Quality Planning and Management, 2003]).

Two separate reports were initially required for every state. Section 305(b) required the biennial submission to congress, through U.S. EPA, of a report that assessed the quality of all of a states' waters, listed all point-source discharges, identified waters in which water-quality standards are not being attained, and identified the major sources of pollution causing the impairments (whether point or nonpoint). These 305(b) reports have been prepared and submitted to U.S. EPA by the states regularly since the initial 1975–1976 submission requirement.

The second reporting requirement was that of Section 303(d) to identify waters not expected to meet water-quality standards even after the application of technology-based effluent limits. Up until the mid-1990s, most states failed to submit 303(d) lists at all, and the ones that were submitted were generally cursory (Houck, 1997a). It was not until the 1996 submission deadline (with many states not actually submitting lists until 1997) that all states managed to comply. U.S. EPA has implemented Section 303(d)'s requirement for updating the lists "from time to time" by requiring revised lists to be submitted every other year. All states submitted lists in 1998, and U.S. EPA waived the 2000 submittal requirement while it developed further guidance on how to structure and prepare the lists.

U.S. EPA (2001a) recommended (but did not require) that states combine their 305(b) and 303(d) lists into an "Integrated Water Quality Monitoring and Assessment Report". The first such integrated lists were submitted by the states in 2002. These reports, still typically known as "lists of impaired waters" or 303(d) lists, now form the starting point for water-quality management activities. The status of its receiving waters on the 303(d) list is of critical importance to a WWTP, as it will affect its discharge requirements and possibly its ability to expand. As with all aspects of water-quality management, determining the attainment status of waters is not without technical, scientific, or legal difficulty.

DATA NEEDS

The first issue that arises in assessing waters is knowing which data should be used. Guidance issued by U.S. EPA for use by the states in preparing their 1994 submissions advised that states should use "...existing readily available data and information and best professional judgment...". Possible data sources include Section 305(b) reports, toxic chemical release inventory data, storage and retrieval (U.S. EPA water-quality database; http://www.epa.gov/STORET/) data, fish consumption advisory information, anecdotal information, and public reports (U.S. EPA, 1993).

No guidance on data quality assurance was issued; as a result, many waters were listed as impaired with little real certainty about their actual status. In addition, most states lacked the monitoring data to fully assess all of their waters. The GAO was asked by congress to address this issue and reported in 2000 that "states collectively assess only a small percentage of waters in the United States...", that only "19% of the nation's rivers and streams were assessed for the 1996 inventory", and "it would be cost-prohibitive to monitor all of the waters in the country" (GAO, 2000).

U.S. EPA and others have also recognized this problem (AMSA 2002; EPA, 2002b; NRC, 2001). U.S. EPA's federal advisory committee on the TMDL program (TMDL Federal Advisory Committee Act [FACA] committee) strongly recommended that states establish quality assurance/quality control programs to ensure that sufficient data were used in listing determinations and that it was of sufficient quality (U.S. EPA, 1998c; p 11). U.S. EPA responded to this recommendation by releasing guidance entitled Consolidated Assessment and Listing Methodology (U.S. EPA, 2002a). This guidance specifically addresses monitoring design, statistical approaches, modeling to determine status, and specific guidance on using chemical, biological, toxicity, bacteria, and habitat data. In its guidance on preparing 2004 303(d) lists, U.S. EPA provided additional guidance on data quantity, sufficiency, representativeness, and quality (U.S. EPA, 2003a).

Over the past 10 years or so, U.S. EPA has steadily tried to improve the scientific rigor of the TMDL program, beginning with data needs. Wastewater treatment plants should be cognizant, however, that problems could still remain and should ascertain the status of their receiving waters and become familiar with the data cited to support the determinations. Chapter 6 of this book deals, in more depth, with data needs and the analytical and modeling needs of water-quality management.

MEASURING IMPAIRMENT

Once all available data are in hand for a particular water body, it is evaluated to determine if the water is achieving its water-quality standards, including both the designated uses and the numeric and narrative criteria. This is not always a straightforward matter, and states must make decisions on how to interpret compliance. For example, does a single exceedance of a numeric criterion constitute a violation of a water-quality standard? If not, how many exceedances in a given data set would constitute an impairment? This problem is especially important when data sets are small because of the increased probability of types I and II statistical errors (see Chapter 6 for a thorough discussion of these issues). U.S. EPA's Consolidated Assessment and Listing Methodology (U.S. EPA, 2002a) addresses the problem in a comprehensive way (e.g., Sec 4.3.2: How Does the State Make Attainment/Impairment Decisions in the Absence of a 'Perfect Data Set'?).

As noted earlier, U.S. EPA also provided additional guidance in 2003 for use by the states in preparing their 2004 303(d) lists (U.S. EPA, 2003a). Among other things, U.S. EPA addressed interpretation of water-quality standards; the need to include frequency of exceedance of a criterion along with magnitude and duration; statistical approaches to determining impairment (and the need for proper statistical procedures, such as determining confidence intervals); and the need for public review and comment on all aspects of the listing process. Most importantly, U.S. EPA stressed the need for states to develop and document assessment methodologies that are consistent with water-quality standards, sound science, and statistics.

If biological criteria, such as species diversity indices, are used to assess in-stream conditions, impairments may be found (or defined) without identification of the stressor causing the impairment. Because TMDLs cannot be developed without identification of the pollutants causing the impairments, additional monitoring and analytical work would be necessary. U.S. EPA's 2003 guidance stated that "states using biological assessments to make assessment determinations should also consider other types of data and information (i.e., chemical and physical)" (U.S. EPA, 2003a).

It is important to note that the 1987 Amendments to the Clean Water Act made a distinction between pollutant ("dredged spoil, solid waste, incinerator residue, biological materials, radioactive materials, heat, wrecked or discarded equipment, rock salt, cellar dirt, and industrial, municipal, and agricultural waste discharged into water") and pollution ("the man-made or man-induced alteration of chemical, physical, biological and radiological integrity of water"). While pollutants are a subset of pollution, Section 303(d) deals only with pollutants; hence, water-quality impairments caused by factors outside of the definition of pollutant (such as streambank erosion) may not be within the jurisdiction of the TMDL program (NRC, 2001). The identification of the stressors then becomes an important legal and technical consideration in preparing 303(d) lists.

If WWTPs discharge to waters listed as impaired, they should become familiar with the methodology used to make those determinations. They should also take every advantage of their opportunities to review and comment on listing methodologies, proposed or in use.

STRUCTURE OF THE IMPAIRED WATERS LIST

In its November 2001 guidance memo (U.S. EPA, 2001a), U.S. EPA recommended that states refine their lists by creating a number of categories of attainment. The categories are as follows:

(1) Attaining the water-quality standards and no use is threatened.

(2) Attaining some of the designated uses; no use is threatened; and insufficient or no data are available to determine if the remaining uses are attained or threatened.

(3) Insufficient or no data and information to determine if any designated use is attained.

(4) Impaired or threatened for one or more designated uses but does not require the development of a TMDL.

(a) TMDL has been completed.

(b) Other pollution control requirements are reasonably expected to result in the attainment of the water-quality standard in the near future.

(c)   Impairment is not caused by a pollutant.

(5)   The water-quality standard is not attained; The hydrologic assessment unit (AU) is impaired or threatened for one or more designated uses by a pollutant(s) and requires a TMDL. (Note that under the 1997 guidance, waters currently meeting standards but not expected to do so in the future should be included in category 5 as "threatened" waters requiring TMDLs.)

In the new integrated water-quality monitoring and assessment report that states may submit to U.S. EPA instead of separate 305(b) reports and 303(d) lists, these five categories replaced what previously constituted the 305(b) report, while category five alone constitutes the 303(d) list. The ability to use this expanded list of categories has enabled the states to produce lists of impaired waters that are more realistically focused on actual water-quality problems and has improved their ability to assign more useful priority rankings.

DEVELOPMENT OF TOTAL MAXIMUM DAILY LOADS

Total maximum daily loads are simple and powerful in their logical structure. U.S. EPA's 1991 guidance on developing TMDLs (U.S. EPA, 1991a) states

> The objective of a TMDL is to allocate allowable loads among different pollutant sources so that the appropriate control actions can be taken and water-quality standards achieved... .

The TMDL determines the allowable loads and provides the basis for establishing or modifying controls on pollutant sources.

Well-done TMDLs constitute a comprehensive watershed-management approach and a vehicle for addressing most facets of water-quality management, from desired uses and attainability to the implementation of solutions, ideally all done with input from an involved public. A TMDL typically contains the following elements:

■ A statement of the water-quality problem;

■ An analysis to determine the level of pollutant loading that would achieve water-quality standards; and

■ An allocation of the allowable load to the various sources in the watershed, both point and nonpoint, including an allowance for future growth and a margin of safety.

The TMDL also may contain a statement of how the TMDL will be implemented.

It is important for a WWTP to understand how each of these four elements are handled in the development of a TMDL affecting its receiving water. As with all water-quality management programs, there are a large number of scientific, technical, legal, and financial issues involved in analysis and decisionmaking. Each of the four TMDL elements is examined in more detail below.

Statement of the Water-Quality Problem

While this is generally a straightforward issue, it must be kept in mind that the causes of an impairment are sometimes not very well understood in the initial stages of a TMDL and may even have been misidentified. There may also be multiple causes. It cannot be automatically assumed that the TMDL is addressing the proper pollutant or the most important cause of the impairment. Wastewater treatment plant discharges can be (and have been) incorrectly identified as causes of impairments.

Analysis to Determine the Allowable Load

In this step, the pollutant loads from the various sources in the watershed (and possibly beyond) are linked to conditions in the water body. Some degree of water-quality modeling is generally required to do this. Analytical techniques, ranging from simple dilution calculations to dynamic, three-dimensional models, are used.

This introduces a host of issues related to data needs, model selection and validity, analytical uncertainty, and prediction reliability. (Chapter 6 discusses these issues in greater detail in the context of a trading program).

The selection of design conditions will have a large effect on the conclusions of the analysis. U.S. EPA's TMDL regulations require that the TMDL account for "seasonal variations" and "critical conditions for streamflow, loading, and water-quality parameters" (40 CFR 130.7 [Water Quality Planning and Management, 2003]). As a result of these requirements, many TMDLs analyze water quality for both annual average conditions and low-flow conditions. While summer average conditions are sometimes used, the seven-day average low-flow that is expected to occur once every 10 years (the 7Q10) is used more often. This is the same conservative low-flow condition generally used by all states to calculate WQBELs when issuing NPDES permits. Hence, its use in TMDL analyses brings the same high degree of conservatism found in the permitting process (e.g., WWTPs discharge at maximum permitted flows and concentrations for all parameters during the critical low-flow conditions).

In the TMDL context, it also raises the issue of the treatment of nonpoint-source loads during critical low-flow conditions. Some TMDLs have made the questionable assumption that there are no nonpoint-source loads during low-stream flows, ignoring groundwater input, in-stream recycling, benthic loads, or loads delivered by short, intense rain events (probably as a result of U.S. EPA's 1991 guidance itself making this assumption).

In the low-flow allocation, then, the full burden of pollutant reduction falls on the point-source dischargers. In water bodies where the predominate sources of the pollutant are nonpoint sources, the average annual and low-flow allocations may be very different because of this.

The analysis should also address the locations of the impairments and discharges and account for any delivery ratios or attenuation affects that would moderate the effect of the WWTPs end-of-pipe discharge.

Allocations

Once the allowable total pollutant loads are determined, they are then allocated to the various sources. The allocation to point sources is referred to as the wasteload allocation, and the allocation to the nonpoint source is referred to as the load allocation (U.S. EPA, 1991a). While Section 303(d) of the CWA used the phrase "total maximum daily load", U.S. EPA regulations state that TMDLs can be expressed in terms of "either mass per unit time, toxicity, or other appropriate measure" (40 CFR 130.2 [Water Quality Planning and Management, 2003]).

U.S. EPA regulations also require the allocations to contain a "margin of safety that takes into account any lack of knowledge concerning the relationship between effluent limitations and water quality" (40 CFR 130.7 [c] [1]). U.S. EPA guidance states that the margin of safety (MOS) "is normally incorporated into the conservative assumptions used to develop TMDLs" or can be added as a separate component (U.S. EPA, 1991a). The NRC study noted that the MOS "is typically an arbitrarily selected numeric safety factor" and that it is sometimes controversial because "it is meant to protect against potential water quality standards violations, but does so at the expense of possibly unnecessary pollution controls" (NRC, 2001; p 74).

The NRC recommended that the MOS "should be determined through a formal uncertainty and error propagation analysis." Thus far, however, not much progress has been made toward this goal.

In summary, a TMDL is defined as follows:

TMDL = Load allocation + Wasteload allocation + MOS

U.S. EPA also recommends allowing for future growth in both point and nonpoint sources, if expected. The projected growth in loads should be included in the load and wasteload allocations and not as separate terms in the equation. The TMDL FACA committee went beyond that and recommended that future growth always be considered in allocations and that the TMDL should contain a discussion of the implications of the allocation for growth (U.S. EPA, 1998c; p 35).

Public Review

Public review of proposed TMDLs does not appear to be absolutely required. U.S. EPA regulations merely state "calculations to establish TMDLs shall be subject to public review as defined in the state CPP [continuing planning process]." U.S. EPA guidance, however, unequivocally states that "states are expected to ensure appropriate public participation in the TMDL development and implementation process" (U.S. EPA, 1991a).

Few, if any, states have attempted to limit public involvement (including WWTP involvement) in the TMDL process. However, many times, the draft TMDL is only made available for public review and comment for a 30-day period, after which further comments are not generally accepted. For TMDLs or water bodies of any complexity, it is difficult for a WWTP, or anyone else, to complete the necessary investigation and technical analysis needed to fully assess the validity of the proposed TMDL within the 30-day comment period. This is one more reason why WWTPs should closely follow 303(d) listing activities and should get involved in the TMDL process at the earliest possible opportunity. Most states would welcome the technical assistance and input from WWTPs and others if it were offered in a cooperative spirit.

Review and Approval

The CWA requires states to submit their TMDLs to U.S. EPA for review and approval. The language actually requires U.S. EPA to approve or disapprove a TMDL within 30 days of its submittal. If U.S. EPA disapproves a TMDL, the Act requires U.S. EPA itself to prepare a TMDL for the impairment, within 30 days of its disapproval of the state's TMDL. U.S. EPA has not been able to comply with these unrealistic deadlines and, in practice, the process takes much longer.

Implementation

Implementation of TMDLs is an interesting issue from several aspects. The language of the CWA does not actually require implementation; it only requires their development. Neither do U.S. EPA's TMDL regulations explicitly require full implementation; the regulations only require that the wasteload allocation be implemented through NPDES permits (40 CFR 130.7 [Water Quality Planning and Management, 2003]). These facts have led some to argue that TMDLs do not have to be implemented; however, these arguments have not been very successful.

The TMDL FACA committee felt strongly that implementation should be fully addressed at the time a TMDL is developed. It recommended that U.S. EPA issue regulations that would require both an implementation plan and schedule to be submitted to U.S. EPA with the TMDL. Further, the FACA committee recommended that the regulations require the implementation plan to contain all of the following elements (U.S. EPA, 1998c; p 37–41):

- Descriptions of actions (control actions and/or management measures) that will be implemented to achieve the TMDL;
- A schedule for implementing specific activities;
- The legal authorities under which the activities will be carried out;
- Reasonable assurances;
- An estimate of the time required to attain applicable water-quality standards, and a demonstration that the standards will be met as expeditiously as practicable;
- A monitoring plan;
- Measurable milestones;
- The ramifications of failing to meet these milestones; and
- A schedule for revising the state's CPP and applicable WQMPs.

The final answer on implementation plan requirements will be known if and when U.S. EPA promulgates new TMDL regulations. U.S. EPA guidance, on the other hand, unequivocally calls for TMDL implementation (U.S. EPA, 1991a).

Also required are "reasonable assurances" that the wasteload and load allocations would be met. The TMDL must also contain a description of the state's implementation plan. For point-source loads, the TMDL is to be implemented through NPDES permits. Permit limits that would collectively achieve the wasteload allocation are to be added for all of the point-source dischargers. Because all NPDES permits issued by the states have to be approved by U.S. EPA, this gives U.S. EPA the power to ensure that TMDL requirements are actually incorporated to permits.

However, the implementation of a TMDL and its wasteload allocation does not necessarily mean that all point sources discharging the impairing substance must have limits placed in their NPDES permits. Under U.S. EPA regulations, permit limits are required only when a discharger has a "reasonable potential" to cause or contribute to the violation of a numeric or narrative water-quality criterion (40 CFR 122.44(d) [National Permit Program, 2004]). The regulations do not stipulate any particular methodology for determining reasonable potential; however, U.S. EPA provided a statistical method in its Technical Support Document for Water Quality Based Toxics Control (U.S. EPA, 1991b).

Implementation is not so easy with nonpoint-source loads, however, because of the central contradiction in water-quality management in the United States today; the CWA does not grant the federal government any direct authority to control nonpoint-source pollution. This responsibility was considered by congress to be solely a state prerogative. Hence, U.S. EPA does not have the authority to require states to actually implement the nonpoint-source control measures that would achieve the load allocations of a TMDL. U.S. EPA guidance has progressed from a general discussion of the need for BMPs (U.S. EPA, 1991a) to "veiled threats" to withhold state grant funds if load allocations are not implemented in some manner (Houck, 1998; p 10420).

What the 1991 U.S. EPA guidance does require regarding non-point-source controls should be of some concern to point-source dischargers. It states

> When establishing permits for point sources in the watershed, the record should show that in the case of any credit for future nonpoint source reductions, (1) there is a reasonable assurance that nonpoint source controls will be implemented and maintained or (2) that nonpoint source reductions are demonstrated through an effective monitoring program (U.S. EPA, 1991a).

The guidance then goes on to state

> it may be appropriate to provide that a permit may be reopened for a wasteload allocation, which requires more stringent limits because attainment of nonpoint-source load allocation was not demonstrated (sic).

This issue is discussed further below under the heading Equity.

> It should be noted that for coastal waters, Section 6217 of the Coastal Zone Act Reauthorization Amendments of 1990 (CZARA) (Coastal Nonpoint Source Pollution, 1990) requires states with approved coastal zone management programs to develop coastal nonpoint pollution control programs. The CZARA also gives U.S. EPA additional significant means to ensure that the management measures are implemented.

Phased Total Maximum Daily Loads

Many, if not most, TMDLs will be relatively straightforward and can move into the implementation stage without undue concern. There are some conditions, however, where a more prudent stepwise approach to the TMDL is warranted. U.S. EPA has termed these "phased" TMDLs (U.S. EPA, 1991a). According to the guidance, conditions warranting a phased approach include the following:

■ Inadequate data to establish allocations;

■ Inadequate predictive tools to assess water-quality response; and

■ Both point and nonpoint sources are involved, and the point source wasteload allocation is based on a load allocation for which nonpoint-source controls need to be implemented.

The 1991 guidance went so far as to suggest that "States may actually prefer it [the phased approach] because the additional data collected can be used to verify expected load reductions, evaluate effectiveness of control measures, and ultimately determine whether a TMDL needs to be revised" (U.S. EPA, 1991a). The states, however, have almost invariably preferred (or been required by a consent decree) to adopt a "get-it-over-quick, the-court-has-given-us-deadlines approach", and phased TMDLs have been relatively rare to date. The legal mandates themselves may sometimes exclude the use of phased TMDLs.

Wastewater treatment plants should be aware, however, that wasteload allocations would be determined and implemented on an interim basis under a phased TMDL. U.S. EPA guidance provides the flexibility that the wasteload allocations could either maintain existing permit limits or establish new ones. The WWTP should try to ensure that any interim limits, imposed in the context of inadequate data and analytical uncertainty, are not unnecessarily or unfairly stringent. The WWTP should also keep in mind, however, that the regulatory agency would be struggling to balance allocation fairness and water-quality protection in the face of inadequate data (and the interim limits may actually be too lax); hence, there should probably be a fairly high threshold for this type of challenge.

Informational Total Maximum Daily Loads

Occasionally, the TMDL analysis may find that the water body is not actually impaired. Rather than completing the TMDL for submission to U.S. EPA for approval, the TMDL could be issued by the state simply for informational purposes or as a watershed management plan (AMSA, 2002). Of course, the water body should be removed from the 303(d) list as well.

ISSUES AND CONCERNS WITH TOTAL MAXIMUM DAILY LOADS

The blossoming of the TMDL program, despite its dating from 1972, is a relatively recent phenomenon. It is a large, complex, and technically demanding program that is somewhat immature at this stage of its development. Because of this, there are a number of TMDL issues of which WWTPs should be aware.

Lack of Historical Success

The first lack of historical success for TMDLS is the approach itself. The TMDLs link pollutant control requirements to ambient water-quality standards, and the history of such water-pollution control programs in the United States has not been good. As Houck (1999) put it , these programs "have always relied more on science than science can deliver... . They require proof of causes and effects that, arguably, come from other causes, and have other effects, and pinning the tail on the right donkey has plagued air, water, and toxics programs from their inception... ." (Houck, 1999; p 10474).

Resource Needs

Ambient-based, water-quality programs are also very resource-intensive. Money and personnel are needed for monitoring and data collection, analysis and modeling, interacting with stakeholders, TMDL preparation, and enforcement. The data needs for water-quality assessments and models are large, and complex water-quality models can be very demanding to properly use. Neither U.S. EPA nor the states have adequate resources for these efforts, and, without adequate resources, the science suffers. When the science of water-quality management suffers, WWTPs have reason to fear being a donkey found in the wrong place at the wrong time.

Scientific Uncertainty

The technical challenges of TMDLs are also daunting. They must identify and quantify myriad sources of pollution and link them to water quality. A range of models, from simple to complex, is used to do this. Unfortunately, as a top water-quality modeler recently observed

> ... most [models] are little different than those developed in the1960s and 1970s for dry-weather wasteload allocation... Worse yet, many models have complex detailed input requirements, so many people assume that the calculations are more precise and accurate than they really are—and little guidance on testing model adequacy or reliability is available (Freedman, 2001).

Perhaps more worrisome is the fact that many of the models currently in use are deceptively simple to run by users who may not fully understand their limitations.

Because of the importance of the issues of data needs, analytical tools, and scientific uncertainty in water-quality management and water-quality trading, they are covered more fully in Chapter 6 of this book.

Nonpoint-Source Pollution

An examination of the nationwide 1998 303(d) list of impaired waters showed that sediment, nutrients, and bacteria were the most prevalent causes of impairment, and that 43% of the listed waters were impaired solely by nonpoint sources, 10% by point sources, and the remaining 47% by a combination of point and nonpoint sources (Freedman, 2001). This finding points directly to the central contradiction of the CWA—it has been extremely successful in controlling point-source pollution, but because it does not regulate most nonpoint-source pollution, "...unregulated sources have blossomed like algae to consume the gains." (Houck, 1999; p 10470).

The causes of nonpoint-source pollution vary widely across the country. Agriculture is a major source. In urban areas, stormwater runoff is the cause of many impairments, particularly bacteria and sediment. Although it is regulated as a point source, there is little immediate prospect of establishing numeric permit limits that would ensure that water-quality standards are met. Additionally, in some areas of the country, particularly the east, atmospheric deposition is an important source of nonpoint-source pollution.

So the fundamental question facing the TMDL program is whether it will be able to overcome this shortcoming. Whether TMDLs can successfully deal with the nonpoint-source problem has serious implications for WWTPs. The first concern is the previously noted fact that the water-quality models typically used by state regulatory agencies are simple steady-state models that were developed to determine dry-weather wasteload allocations for WWTPs. Such modeling does not adequately deal with nonpoint-source pollution, which requires dynamic modeling, a much more data and resource-intensive proposition. (One interesting illustration of this is the finding that air deposition is turning out to be an important source of some pollutants, notably nutrients and mercury. However, few states are able to link air and water-quality modeling in the TMDL context).

Some states simply apply their steady-state models to both low-flow and average flow conditions and produce load allocations for both, neither of which capture the transient, wet-weather effects of nonpoint-source loads. Many times, the result is a TMDL with very tight wasteload allocations for the point sources, and marginal or meaningless load allocations for the nonpoint sources, even in cases where 70, 80, or even 90% of the annual pollutant load is from nonpoint sources. As one WWTP recently observed of its state regulatory agency, "When your only tool is a hammer, all your problems look like nails. But this TMDL needed a screwdriver" (Anonymous).

Unfortunately, the fact that TMDL writers essentially have the power to regulate point sources, but not nonpoint sources, also makes WWTPs look like large, inviting nails. In fact, U.S. EPA guidance suggests that states take exactly that approach. If reasonable assurances for the successful implementation of nonpoint-source load allocations are not available, the guidance calls for reducing the point-source wasteload allocations even further (U.S. EPA, 1991a). It is also true that U.S. EPA has continually applied pressure to the states to include meaningful load allocations in all TMDLs and provide ways to provide reasonable assurance.

Allocation Methods

Most TMDLs prepared, to date, have simply approached allocating the allowable load as an ad hoc mathematical exercise influenced mainly by the perceived performance capabilities of WWTPs and, somewhat, by the availability of reasonable assurances for nonpoint-source controls. The NRC, in its TMDL study, recognized that the allocation step is far more than that: "Allocation is first and foremost a policy decision on how to distribute costs among different stakeholders in order to achieve a water quality goal" (NRC, 2001; p 98). Hence, the role of science in the allocation process is to simply identify when different pollution control actions are equivalent (i.e., produce the same water-quality response and have the same degree of uncertainty). Building on this insight, U.S. EPA's Twenty Needs Study (U.S. EPA, 2002c) identified the need to develop methods to integrate technical, social, and economic factors in the setting of allocations. The TMDL FACA committee recommended that U.S. EPA distribute "informational guidance" on allocation methods that have been successfully used (U.S. EPA, 1998c; p 36). Until all of this is fully realized, however, the allocation process will probably remain largely an exercise in "ratcheting down" on point sources.

Equity

"Ratcheting down on point sources" unfortunately is a phrase heard too frequently in conjunction with TMDLs. In 1998, Houck examined 55 approved TMDLs that he obtained from U.S. EPA regions across the country. After reviewing them, his conclusions regarding meaningful reductions of nonpoint-source loads are not reassuring for WWTPs. He states "Nonpoint sources are targets of last resort" (Houck, 1998; p 10437); and

> Even where identified, nonpoint-source reductions will frequently not be calculated, and, where calculated, even more frequently will not be implemented through identified abatement plans (Houck, 1998; p 10437).

Moreover, the NRC TMDL study, citing a National Academy of Public Administration report (NAPA, 2000) found that

> In some cases, point source permitting is used to impose conditions on point sources that essentially requires them to finance control practices for unregulated nonpoint sources (NAPA, 2000; p 86).

The NRC report recognized that WWTP resistance to these inequities would probably be manifested as technical critiques of the TMDL analysis itself, and more importantly, it stressed that

> Distributing the cost and regulatory burdens for designated use attainment in a way that is deemed equitable by all stakeholders is critical to future TMDL program success (NRC, 2001; p 100).

The TMDL FACA committee agreed that allocations should be equitable, but also stressed that they must achieve water-quality goals (U.S. EPA, 1998c; p 35).

ADAPTIVE IMPLEMENTATION

It is clear that water-quality management is accompanied, every step of the way, by uncertainty. Some stakeholder groups feel that this uncertainty justifies the use of very conservative and strict regulatory approaches. Others, notably the regulated community, call for more "sound science", that is, less uncertainty, in the setting of regulatory requirements, lest scarce resources be wasted in implementing needless or ineffective requirements. At their extremes, neither of these two opposing philosophies are very useful in attempting to improve water-quality in the real world of resource limitations and unavoidable uncertainty.

A more pragmatic, middle-ground approach to the TMDL has been proposed. The TMDL FACA committee recommended that TMDLs contain provisions for follow-up monitoring, evaluation, and potential revision, to "allow for an iterative (or adaptive or phased) approach in cases of uncertainty or lack of success in achieving standards" (U.S. EPA, 1998c; p 43–45). Freedman (2001) termed this concept "adaptive watershed management" and described it as

> Using the best tools and data available, we should make best estimates and take action, recognizing that the decision and action may not be final. If we work to explicitly define the range of uncertainty in our analysis, we can act within that range. Then if, as part of the TMDL, we monitor progress and later adapt our actions, we can continue to progress toward clean water (Freedman, 2001).

The NRC (2001) termed it "adaptive implementation" and considered it nothing less than the incorporation of the scientific method into the TMDL process.

> It is a process of taking actions of limited scope commensurate with available data and information to continuously improve our understanding of a problem and its solutions, while at the same time making progress toward attaining a water quality standard. Plans for future regulatory rules and public spending should be tentative commitments subject to revision as we learn how the system responds to actions taken early on (NRC, 2001; p 90).

Adaptive implementation in the TMDL program was a fundamental recommendation of the NRC.

In practical terms, the adaptive approach to TMDLs and water-quality management, in general, means that things don't have to be perfect to proceed. Nor should far-reaching or expensive requirements be mandated in the face of excessive uncertainty. The key is to work continuously to improve scientific understanding as steady progress is made toward water-quality goals. Perhaps to the occasional chagrin of WWTPs, the most certain and often fastest progress can be made through point-source reductions.

The Long Island Sound nitrogen TMDL is an excellent example of how to implement the adaptive management approach (Stacey, 2004). Connecticut, New York, and U.S. EPA recognized that much could change over the 15-year implementation period; thus, a provision for review and reissuance was included so that new technologies or criteria could be incorporated and adjustments made if insufficient progress were being made.

Trading and Water-Quality Management and Planning

Is water-quality trading consistent with the CWA and all of its relevant requirements? Or, as asked in the opening paragraphs of this chapter, Is water-quality trading legal? What would be the role of trading in achieving water-quality goals? How does it relate to each element of water-quality management—planning, water-quality standards, TMDLs and load allocations, and permitting? The remainder of this chapter address these questions and assesses the consistency of trading with the CWA, its implementing regulations, and each element of water-quality planning and management.

AUTHORITY TO TRADE

U.S. EPA's Water Quality Trading Policy (U.S. EPA, 2003b) repeatedly states that water-quality trading programs must be consistent with the CWA and the nation's existing water-quality programs.

> Clear legal authority and mechanisms are necessary for trading to occur... Provisions for water quality trading should be aligned with and incorporated to core water quality programs (U.S. EPA, 2003b).

The policy goes on to state that U.S. EPA believes that trading is consistent with the CWA and all associated regulatory requirements.

> [U.S.] EPA believes the CWA provides authority for [U.S.] EPA, states and tribes to develop a variety of programs and activities to control pollution, including trading programs...[U.S.] EPA believes this may be done by including provisions for trading in water-quality management plans, the continuing planning process, watershed plans, water quality standards, including antidegradation policy and, by incorporating provisions for trading into TMDLs and NPDES permits (U.S. EPA, 2003b).

U.S. EPA clearly regards trading as merely another tool, albeit a promising but largely untried one, to be added to the CWA-based tool box. It believes that the existing statutory and regulatory programs provide the legal basis and flexibility to incorporate trading.

Not everyone agrees. In an April 2002 letter to U.S. EPA's assistant administrator for water, a coalition of environmental organizations characterized trading as potentially an unraveling of CWA regulatory programs and further claimed that the CWA provides no authority for trading programs.

> The CWA does not contain any mechanism by which point source dischargers can achieve their obligations to meet water-quality based effluent limitations or technology standards by trading with other point or nonpoint sources (Chesapeake Bay Foundation et al., 2002).

In the view of these organizations, water-quality trading would have to be specifically authorized by the CWA before trading programs were developed. It appears that these organizations perhaps would find trading to be consistent with the CWA if its sole purpose were offsetting future growth in loads. This is a form of the issue of "trading to achieve versus trading to maintain," which frequently arises in the development of trading programs and is discussed in more detail below. Despite the strength of the statements in this letter, only one legal challenge to trading has been mounted to date, and it was unsuccessful (see the discussion of Ohio Valley Environmental Coalition et al. versus Whitman in the Antidegradation section later in this chapter).

WATER-QUALITY PLANNING

The CWA established a structure for state water-quality planning activities and U.S. EPA oversight and approval. The most important element of the planning activities is the CPP required by Section 303(e). States are required to continuously update their water-quality planning activities to stay focused on the most critical current water-quality problems. According to U.S. EPA regulation (40 CFR 130.5 [Water Quality Planning and Management, 2003]), a state's CPP must include, among other things, descriptions of the following:

- The state's process for developing effluent limitations and schedules of compliance;

- The state's process for incorporating elements of area-wide waste management plans and applicable basin plans;

- The state's process for developing TMDLs and individual water-quality based effluent limits;

- The state's procedures for updating and maintaining WQMPs;

- The state's process for establishing and insuring adequate implementation of new or revised water-quality standards;

- The state's process for developing an inventory and priority ranking of needs for construction of WWTPs; and

- The state's process for determining the priority of permit issuance.

It can be seen from this list that the CPP is the means by which states identify the methods and processes to be used to manage water quality. It is also the means by which U.S. EPA tracks the elements of a state's management processes.

The WQMPs included in the above list are the other critical component of the planning process. Whereas the CPP sets out programmatic processes, WQMPs are used to direct implementation of solutions to specific water-quality problems. U.S. EPA regulations define them succinctly.

> Water Quality Management Plans draw on water-quality assessments to identify priority point and nonpoint source water quality problems, consider alternative solutions and recommend control measures, including financial and institutional measures necessary for implementing solutions (40 CFR 130.6 [Water Quality Planning and Management, 2003]).

The regulations require that the state's actual work program be based on the priorities established in the WQMP. Elements of the WQMP include the following:

- Total maximum daily loads;
- Effluent limitations and schedules of compliance;
- Identification of anticipated WWTPs;
- Establishment of construction priorities and schedules;
- Nonpoint-source management and control (for both regulatory and nonregulatory programs);
- Identification of responsible management agencies; and
- Identification of implementation measures necessary to carry out the plan.

One of the simplest mechanism for providing "clear legal authority for trading" is for a state to include a description of its trading programs in its CPP and specific applications of trading in its WQMPs. There is nothing in the statutory or regulatory requirements that precludes the inclusion of trading as a water-quality management tool, and its inclusion in CPPs and WQMPs provides the legal authority for trading and the integration of trading into core water-quality management programs sought by U.S. EPA.

Some states may wish to promulgate regulations to authorize and regulate trading, as Michigan did for trading anywhere in the state (Michigan Department of Environmental Quality, 2002) and North Carolina did for the Neuse River Compliance Association. Connecticut developed and passed enabling legislation for the Long Island Sound nitrogen trading program.

WATER-QUALITY STANDARDS

As described in the first part of this chapter, water-quality standards are the goals that the water-quality management process are mandated to attain. The standards themselves give no specifics on how they are to be achieved. Hence, there is no conflict between the establishment of water-quality standards and the creation of trading programs as one of the methods developed to help achieve them.

Like all water-quality management tools, the purpose of trading is to help achieve water-quality standards, and it cannot become an instrument for violating them. A discharge permit could not be issued if it were to allow a discharge that would result in violations; neither could a trading program be authorized that would do so. Hence, trading to help achieve a water-quality standard at one location in a water body cannot result in violations of standards at another location.

ANTIDEGRADATION

In the development of any trading program, considerable consideration should be given to how a state's antidegradation provisions would be applied. While antidegradation should not generally be a problem for the reasons discussed below, it could place restrictions or outright prohibitions on water-quality trading.

Antidegradation should not be an issue for trades involving tier I waters. Because NPDES permits that would result in violations of existing water-quality standards cannot be issued, and trades would be incorporated to permits (either directly or indirectly [see Chapter 5]), a trade simply could not result in a violation of a water-quality standard; if it did, it would mean that the permit was improperly issued.

However, antidegradation could come into play with trades involving new or increased discharges to tier II and tier III waters. If a trade would fully offset any increase in loads, it could be argued that antidegradation should not apply and, therefore, antidegradation review should be not required, simply because there would be no net increase in loading. However, to ensure that outcome in the case of a WWTP acquiring credits from a nonpoint source, the WWTP may need to find ways to provide reasonable assurances that there would, in fact, be no net increase in loadings.

Even when a trade would increase loadings of a pollutant to a tier II water, antidegradation could still be satisfied with the appropriate socioeconomic showing. Such a showing could be worked into the permit issuance or modification process to fulfill the public participation requirements associated with the tier II review.

An example of the application of antidegradation to a trade involving a tier II water might be as follows: Assume a WWTP discharges 4.5 kg (10 lb) of copper per day, and the water-quality standards would allow the discharge of 6.8 kg (15 lb). However, the WWTP would like to expand and discharge 9.1 kg (20 lb) per day and proposes a trade to offset 2.3 kg (5 lb) of that, leaving the net loading at 6.8 kg (15 lb) per day. The WWTP would still have to make a tier II antidegradation showing that justified the increase in loading from 4.5 to 6.8 kg (10 to 15 lb) per day.

At least one federal court has ruled on whether trading can be used to minimize or offset increased loadings as a way to satisfy or avoid antidegradation requirements.

In January 2002, a coalition of environmental groups filed a suit challenging U.S. EPA's approval of West Virginia's antidegradation policy, charging that the approval violated the requirements of the CWA (*Ohio Valley Environmental Coalition et al. v. Whitman*, Civ. No. 3:02-CV-59, January 23, 2002). The plaintiffs challenged 13 different aspects of U.S. EPA's approval of West Virginia's antidegradation implementation procedures regulation. Included in the challenged provisions was an express authorization in the West Virginia rules that allowed trading to minimize or offset loadings to avoid or satisfy antidegradation requirements. U.S. District Court Judge Joseph Goodwin issued a ruling on the case in August 2003. Significantly for water-quality trading (Calamita, 2004), the court upheld the provision allowing "new or expanding dischargers to offset pollutants to avoid antidegradation review or to minimize degradation through point or nonpoint source trading".

In summary, antidegradation is an important issue that should be addressed in the early stages of developing trading proposals. For the most part, however, antidegradation should not be a significant obstacle to well-conceived trading proposals. Nevertheless, groups opposed to public and private dischargers having the option of water-quality trading could try using antidegradation to prevent it. Judge Goodwin's decision provides a good early test of using antidegradation for that purpose.

ANTIBACKSLIDING

Concerns have been raised that antibacksliding could interfere with trading by preventing a WWTP from increasing its discharge through the use of credits. In general, however, antibacksliding should not be an obstacle to water-quality trading programs.

The applicability of antibacksliding would depend on the status of the receiving water. As noted earlier in this chapter, the prohibition against relaxing permitted effluent limits would not apply if the new limits were in compliance with a TMDL wasteload allocation. Hence, antibacksliding constraints would not apply to trades involving impaired or restored water bodies where TMDLs have been implemented.

If there were no TMDLs, and the receiving water were in attainment with water-quality standards, then permit limits could still be increased to accommodate a trade, but only if doing so would be consistent with antidegradation policy (Batchelor, 2004).

The trading policy addresses the issue directly, as follows:

[U.S.] EPA believes that the antibacksliding provisions of Section 303(d)(4) of the Clean Water Act will generally be satisfied where a point source increases its discharge through the use of credits in accordance with alternate or variable water quality based effluent limitations contained in an NPDES permit, in a manner consistent with provisions for trading under a TMDL, or consistent with the provisions for pre-TMDL trading included in a watershed plan (U.S. EPA, 2003b). (Note: The citation in the policy is in error; antibacksliding is addressed in Section 402[o].)

While U.S. EPA has good reason to believe that antibacksliding would not prevent increasing discharges through the use of credits, nor presumably prevent the cessation of credit generation by the supplying WWTP, it is still possible that groups opposed to trading programs could mount legal challenges over this issue.

TOTAL MAXIMUM DAILY LOADS

Total maximum daily loads are both analytical frameworks in which water-quality management decisions can be made and allocation decisions that guide regulatory and voluntary pollution control actions. As analytical frameworks, they have the advantage of being a comprehensive watershed approach to analyzing water-quality problems and evaluating possible solutions. Once pollutant loads from all sources are ascertained, the allowable total load necessary to achieve water-quality standards is determined and then allocated to point and nonpoint sources. The point-source allocation is also subdivided to individual point sources as wasteload allocations.

The total maximum allowable pollutant load sets an overall cap for all sources, and the assignment of wasteload allocations to point sources creates the opportunity to exchange discharge allowances. Hence, the TMDL context can be a good framework for water-quality trading. U.S. EPA recognized this in its 1991 TMDL guidance. It described the potential for trade-offs between point sources with similar discharges and between point and nonpoint sources and acknowledges that these trade-offs may be a way to achieve the TMDL goals in a more cost-effective way.

OTHER ANALYTICAL FRAMEWORKS

While TMDLs may provide a strong framework for trading to occur, other frameworks are possible. Trading can occur in watersheds that already meet water-quality standards (and hence no TMDL is required). For example, increased point-source loadings because of growth could be offset by upstream, nonpoint-source reductions implemented by the point source. In this case, modeling of stream transport mechanisms upstream of the point source and water quality downstream of it would be needed to determine the necessary levels of upstream load reductions.

The Draft Framework for Watershed-Based Trading (U.S. EPA, 1996) espoused a principle that, "Trades are developed within a TMDL or other equivalent analytical and management framework." The only requirement for the other frameworks is that they be able to "link pollutant contributions from sources to ambient conditions, and predict the effects of pollutant reductions from different sources on in-stream water quality" (U.S. EPA, 1996).

Of course, trading programs are only possible in water bodies with multiple sources of pollution—either point-source dischargers or nonpoint-source loads that can be estimated with reasonable accuracy. Whether the analytical framework used to set the effluent limits for the point source(s) is a TMDL or something else, it must include the pollutant loads from all sources so that the collective effect does not result in violations of the water-quality standards. A holistic analysis is required.

GEOGRAPHIC TRADING AREAS

In general, geographic trading areas should coincide with the geographic boundaries of the analytical framework, i.e., the watershed or TMDL boundaries (U.S. EPA, 1996, 2003b). While a trading program obviously could not allow trading beyond the boundaries of the watershed, it could restrict trading to a geographic area smaller than the watershed or could create groups of potential trading partners within subwatersheds. The only hard and fast rule is that the traded pollutant reductions must exert a benefit in the areas of concern in the receiving water (i.e., the areas for which the load allocations or permit limits were established). In some cases, such as hypoxia in the Gulf of Mexico or the Chesapeake Bay, this could result in very large trading areas.

There are any number of reasons why states may feel they want to restrict the trading boundaries. In some cases, restricting the geographic scope could increase the viability of trading. If trading partners are far-flung or in different subwatersheds, the analytical needs to ensure that the trades achieve water-quality standards everywhere would be more demanding, and larger trading ratios could be required (see Chapter 5 for a discussion of trading ratios). Likewise, if multiple political jurisdictions or management agencies are involved, the trading program itself becomes more administratively complex. To date, no interstate trading programs have been developed, although interstate trading within the Potomac River basin is being contemplated in the context of the Chesapeake Bay Program (U.S. EPA, 2001b).

In larger watersheds, such as the Chesapeake Bay or Long Island Sound, a tradeoff is involved. The larger the geographic scope, the more potential trading partners there are, so restricting the trading areas for the reasons described above could diminish the viability and cost-effectiveness of trading.

It must be noted that many environmental groups are strongly opposed to trading across watershed (or even subwatershed) boundaries. They tend to see it as "importing" pollution into one watershed to benefit another, thereby creating a "fairness" issue in the receiving location. The program requirement that trades not cause local water-quality impairments does not seem to alleviate this concern.

NATIONAL POLLUTANT DISCHARGE ELIMINATION SYSTEM PERMITS

The trading policy sets a simple prerequisite that relates trading to NPDES discharge permits.

> Sources and activities that are required to obtain a federal permit pursuant to Sections 402 or 404 of the CWA must do so to participate in a trade or trading program (U.S. EPA, 2003b).

> Section 402 (CWA, 1972) addresses point-source dischargers and is applicable to virtually all municipal and industrial dischargers; Section 404 deals with dredge and fill materials. Hence, the policy simply states that a WWTP must have a NPDES permit before it can trade.

> The effluent limits contained in NPDES permits are established within the analytical framework and management context the state has chosen for the receiving water body. Trading can be incorporated to these elements, including the permit, as long as it advances, or at least does not prevent the attainment of the water-quality goals. Hence, as with other areas of water-quality management, there is no obvious inherent conflict between trading programs and NPDES permits.

> The various options for relating trades to NPDES permits and the various issues related to doing so are discussed in Chapter 5.

TRADABLE POLLUTANTS

Some types of pollutants are more suitable for trading than others. An ideal tradable pollutant would have the following characteristics:

- It would have numerous sources and be widespread;
- It would have mainly far-field effects;
- It would be easily measurable;
- A unit of trade could be easily derived; and
- It would not exert toxic effects or be bioaccumulative.

Nationwide, the large majority of trading projects and feasibility studies thus far have been for nutrients. At least one BOD trade has taken place, and a small number of studies have examined metals. The trading policy (U.S. EPA, 2003b) states that U.S. EPA supports nitrogen, phosphorus, and sediment trading, and that the agency recognizes that "trading of pollutants other than nutrients and sediments has the potential to improve water quality and achieve ancillary environmental benefits...". The policy goes on to state, however, that "such trades may pose a higher level of risk and should receive a higher level of scrutiny so that they are consistent with water quality standards" (U.S. EPA, 2003b), and that U.S. EPA may support such trades if there has been prior approval of some sort from U.S. EPA (an NPDES permit, TMDL, watershed plan, etc.).

The first trade involving BOD_5 was the Rahr Malting permit in Minnesota, and it also incorporated cross-pollutant trading (see the Rahr Malting Company section in Chapter 2 for a description of this trade). The Minnesota Department of Environmental Protection quantified equivalency ratios between upstream nitrogen, phosphorus, and BOD reductions, based on their effects on in-stream dissolved oxygen levels downstream of the Rahr discharge. The trading policy states that U.S. EPA supports this type of cross-pollutant trading for pollutants affecting dissolved oxygen where this type of equivalency ratio can be adequately quantified, but the policy does not extend this support to other pollutants.

The trading policy does not rule out the possibility of trades involving bioaccumulative toxics, perhaps because it is conceivable that, in some circumstances (such as the predominate sources being nonpoint), trading may be one way to make progress toward reductions. Perhaps also influencing U.S. EPA is the fact that several feasibility studies or pilot projects have been underway dealing with metals. U.S. EPA's support is qualified, however, and limited to a willingness to consider "a limited number of pilot projects...to obtain more information" (U.S. EPA, 2003b).

The trading policy is silent on trading of nonbioaccumulative toxics, such as ammonia. It does, however, contain a statement that neither an acute aquatic-life criterion nor a chronic aquatic-life or human-health criterion could be exceeded in a mixing zone as the result of a trade. This could be interpreted as allowing for the trading of nonbioaccumulative toxics. However, these types of toxics are generally short-lived and exert only near-field effects; hence, it is difficult to envision many circumstances in which they could be suitable and appropriate for trading.

TYPES OF TRADING

The trading policy discusses the following four possible types of trading:

- Trading of technology-based effluent limits;
- Pretreatment trading;
- Intraplant trading; and
- Trading to comply with water-quality-based effluent limits.

Only the last of these types of trading has potential widespread applicability. The other three are either limited to very narrow circumstances or do not really constitute watershed-based trading.

As described earlier in this chapter, technology-based requirements for WWTPs are defined as conventional secondary treatment levels of BOD_5 and TSS (30 mg/L for both, and 85% removal). Beginning with the 1996 Draft Framework for Watershed-Based Trading and continuing through the 2003 trading policy, U.S. EPA has consistently stated that trading of technology-based requirements would not be supported "except as expressly authorized by federal regulations" (U.S. EPA, 2003b). This last statement refers to certain exceptions for the iron and steel industry currently allowed by U.S. EPA regulations (40 CFR 420.03 [Iron and Steel, 2003]) and would preclude trading of the secondary treatment requirements placed on WWTPs.

While it is clear that U.S. EPA has no intention of expanding these general exceptions to other existing technology-based requirements, the trading policy states that "[U.S.] EPA will consider including provisions for trading in the development of new and revised technology-based effluent guidelines" (U.S. EPA, 2003b). This statement indicates U.S. EPA's willingness to consider incorporating trading provisions into the development of future effluent guidelines (Hall, 2004).

Pretreatment trading is not really watershed-based trading, because it does not deal with discharges to surface waters. Pretreatment programs run by municipalities or wastewater agencies set limits on the levels of certain substances, mainly toxics, that industries may discharge to the municipal sewer system. This is done by first determining the total allowable concentration of a particular substance in the plant's influent flow. The allowable level is designed to prevent the substance from interfering with the WWTP's treatment processes, causing damage to the collection system or plant facilities, or passing through the plant in sufficient quantities to result in violations of water-quality standards in the receiving water. Once determined, the allowable load is then allocated by the pretreatment program in some manner to the various industrial dischargers; the individual allocations are known as local limits. Pretreatment programs currently have a great deal of flexibility in how they set local limits, and there is nothing in the current regulations that prevent voluntary trade-offs between individual industrial dischargers in the setting of these local limits, as long as they discharge to the same WWTP. This book does not address this type of "trading".

Regarding intraplant trading, the trading policy states that "[U.S.] EPA supports intra-plant trading that involves the generation and use of credits between multiple outfalls that discharge to the same receiving water from a single facility that has been issued an NPDES permit" (U.S. EPA, 2003b). Because most WWTPs have only a single outfall (or sometimes a second outfall for peak excess wet-weather flow), intraplant trading will be of limited interest to them. Some industrial dischargers, however, operate two or more outfalls at a given facility that may or may not be in close proximity, though they are covered by the same permit. The statement in the trading policy might seem to suggest that trade-offs between these outfalls would have to be done under the auspices of a trading program and not simply in the context of negotiating the permit limits for the two outfalls. This was not the intent, however; intraplant trading can be done within the NPDES permit without the need for a broader watershed trading program (Hall, 2004).

Most trading programs of interest to WWTPs will involve water-quality-based effluent limits, those limits set according to the levels necessary to achieve water-quality standards in the receiving waters. The impetus for such trading programs and the issues and details they must address will vary according to the status of the receiving waters whose water-quality standards must be met. Requirements will be different, for example, in a watershed where the water body is impaired and a TMDL in place, as compared to one with unimpaired waters and no TMDLs. This is discussed in the following section.

Watershed Scenarios

This section discusses trading and water-quality-based effluent limits in the context of three different watershed scenarios—the receiving water body is unimpaired or unassessed; the water body is on the list of impaired waters, but a TMDL has not yet been completed; and the water body is impaired and a TMDL is in place.

THE RECEIVING WATER BODY IS UNIMPAIRED OR HAS NOT BEEN ASSESSED

While watershed-based trading involving unimpaired waters may not be as common as watershed-based trading involving impaired waters, a number of circumstances can be envisioned that could make trading of interest in such waters. For example, the receiving water is unimpaired, but one of the following conditions apply:

■   A WWTP determines that it would be more economical to meet an existing permit limit for BOD_5 by relaxing the treatment level at the plant, while trading with nonpoint sources or installing BMPs upstream of the plant.

■ A WWTP wants to undertake an expansion that would increase phosphorus loads to the receiving water and result in violation of a water-quality criterion (e.g., phosphorus, dissolved oxygen, or a narrative criterion). Trading is used to offset the increased load.

■ A WWTP agrees to meet a voluntary cap or goal to maintain current water quality and uses trading as one method of complying with the cap.

■ A water-quality standard is made more stringent or a criterion is adopted for a previously unregulated pollutant, and the WWTP would have difficulty meeting the new criterion with its existing processes. If there are other sources of the pollutant in the watershed, either point or nonpoint, the WWTP could use trading to comply with the standard.

As with trading under other circumstances, the goals would be the same—achieving water-quality standards, lowering costs, allowing for growth, etc. The prerequisites for trading described above (e.g., legal authority and analytical framework) would also be the same.

In unimpaired waters, there would be no wasteload or load allocations to use to establish baselines for trading. U.S. EPA's trading policy says simply

> For trades that occur where water quality fully supports designated uses, or in impaired waters prior to a TMDL being established, the baseline for point sources should be established by the applicable water quality based effluent limitation, a quantified performance requirement or a management practice derived from water quality standards (U.S. EPA, 2003b).

In other words, the trade must have the same net water-quality effect as the effluent limit affected by the trade. For a WWTP to sell 4.5 kg (10 lb) of phosphorus credits to another WWTP, it would have to reduce its phosphorus discharge to 4.5 kg (10 lb) less than the mass-loading limit specified in its permit.

While trading in unimpaired waters should be no more conceptually complex than trading under a TMDL, it is possible that antidegradation issues could arise, especially if the water body is a tier II high-quality water. In such cases, care must be taken to ensure consistency with the state's antidegradation policy. While potentially more demanding, there is no reason that trading cannot occur in most, if not all, high-quality waters (see the discussion of antidegradation earlier in this chapter).

Water bodies that are unassessed are a special case and are difficult to generalize. The unassessed status means that there is insufficient water-quality data to determine if water-quality standards are being met. While it is unlikely that any water body receiving permitted point-source discharges would be completely unassessed, it is possible that some portions of the watershed potentially affected by proposed trades have not been assessed. In unassessed or partially assessed watersheds, it is unlikely that the full analytical framework necessary to support a trading program exists, and it would have to be developed before the trading program could be implemented. It is likely that the state would expect the WWTP to share in the costs of data gathering and model development, so the cost of developing a trading program in this type of watershed would likely be higher than in fully assessed ones.

THE RECEIVING WATER BODY IS IMPAIRED, BUT A TOTAL MAXIMUM DAILY LOAD HAS NOT YET BEEN IMPLEMENTED

At first glance, water bodies that fall into this category may not seem to be suitable candidates for water-quality trading programs simply because the necessary analytical framework is unlikely to exist. How can trading programs be designed in the absence of both rules (a wasteload allocation, for example) and tools (a watershed-wide analytical framework)?

While these are valid concerns, the trading policy recognizes that trading could, at times, be beneficial in such water bodies. It states that "[U.S.] EPA supports pre-TMDL trading in impaired waters to achieve progress towards or the attainment of water-quality standards" (U.S. EPA, 2003b). The trading policy goes on to cite three ways in which trading could be beneficial in these circumstances

■ Individual trades achieve a net reduction of the impairing pollutant;

■ Watershed-scale trading reduces loadings to a specified cap supported by some kind of baseline information (possibly an interim goal). Phase I of the Tar-Pamlico trading program (described in Chapter 2) is an example of this type of trading; or

■ Trading that achieves a direct environmental benefit relevant to the conditions or causes of impairment and achieves progress towards restoring designated uses, not necessarily through reducing the loading of a pollutant. A trade that results in multiple benefits, such as streambank restoration to offset a phosphorus discharge, would be an example of this type of trading.

In other words, trading in impaired, pre-TMDL waters essentially means meeting one of three simple criteria: achieving a local reduction of the impairing pollutant; achieving a watershed-wide reduction (this would require some sort of watershed-wide analytical framework); or achieving progress toward restoring the designated use through some other improvement.

The reasons why a WWTP might wish to trade are similar to those described above for unimpaired waters: to meet a permit requirement at lower cost, offset an increased discharge, or meet a voluntary interim cap. The prerequisites for trading would also be the same, except that the analytical framework would be replaced by a demonstration that the proposed trade would result in a net load reduction or would achieve related environmental benefits.

The WWTP should be aware, however, that this type of trading is likely to generate public misunderstanding because it is too easily misinterpreted as making a bad situation worse. Other stakeholder groups may oppose it for this reason. Special care must be taken in the public participation process to show that the proposed trades, in fact, result in net benefits and not further harm. A position that has frequently been asserted by some stakeholders is that trading should not be allowed until all water-quality goals have been met. This was a significant issue in the stakeholder negotiations that produced Chesapeake Bay Nutrient Trading Fundamental Principles and Guidelines (U.S. EPA, 2001b). In this view, the only role for trading would be as a tool to help protect the restored water body sometime in the future.

The WWTP should also be aware that the ultimate requirements of the trading program will depend on the TMDL, once it is developed and adopted. There are no guarantees that requirements imposed or trades approved before TMDL development will remain unchanged once the TMDL is developed. Wastewater treatment plants should avoid staking their futures on interim trading programs but should anticipate changed requirements when the TMDL is adopted.

THE RECEIVING WATER BODY IS IMPAIRED AND A TOTAL MAXIMUM DAILY LOAD HAS BEEN IMPLEMENTED

A TMDL for an impaired water body provides many functions. It is an analytical framework for assessing the problem and identifying solutions. It assigns an aggregate allocation to point sources and a load allocation to nonpoint sources. It can also specify wasteload allocations for individual dischargers, or this may be done subsequent to the adoption of the TMDL. By their very nature, these allocations are tradable discharge allowances. Whether formalized in the TMDL development stage or later, the TMDL provides most of the statutory and regulatory prerequisites for water-quality trading and is the ideal logical framework for it. The only requirement is that all of the water-quality-based effluent limits that are imposed as a result of the TMDL must be consistent with the achievement of water-quality standards per section 301(b)(1)(C) of the CWA (1972). The adjustment of discharges to incorporate trades does not alter this requirement, and the conformance of trades to the requirements provides the legal basis for trading.

References

Association of Metropolitan Sewerage Authorities (2002) *Legal Perspective on Antidegradation*. Washington, D. C.

Batchelor, D., U.S. Environmental Protection Agency, Washington, D.C. (2004) Personal communication.

Calamita, P., AquaLaw PLC Water and Wastewater Solutions, Richmond, Virginia. (2004) Personal communication.

Chesapeake Bay Foundation, Clean Water Action, Coast Alliance, Natural Resources Defense Council, Ocean Conservancy, Sierra Club (2002) Letter to G. Tracy Mehan III, Asst. Administrator for Water, U.S. Environmental Protection Agency, April 18, 2002.

Clean Water Act (1972) U.S. Code, Section 1251–1387, Title 33.

Coastal Nonpoint Source Pollution Control Program (1990) Coastal Zone Act Reauthorization Amendments, Section 6217.

Designated Uses (1995) Code of Maryland Regulations, Subpart 26.08.02.02; Title 26, Department of the Environment.

EPA Administered Permit Programs (2004) The National Pollutant Discharge Elimination System. Code of Federal Regulations, Part 122, Title 40.

Freedman, P. (2001) The CWA's New Clothes. *Water Environ. Technol.*, **13** (6), 28–32.

General Accounting Office (2000) Water Quality: Key EPA and State Decisions Limited by Inconsistent and Incomplete Data; GAO/RCED-00-54. Report to the Chairman, Subcommittee on Water Resources and Environment, Committee on Transportation and Infrastructure, House of Representatives; U.S. General Accounting Office: Washington, D.C.

Government Accounting Office (2003) Water Quality: Improved EPA Guidance and Support Can Help States Develop Standards That Better Target Cleanup Efforts. Report to the Chairman, Subcommittee on Water Resources and Environment, Committee on Transportation and Infrastructure, House of Representatives; U.S. General Accounting Office: Washington, D.C.

Hall, L., U.S. Environmental Protection Agency, Washington, D.C. (2004) Personal communication.

Houck, O. (1997a) TMDLs: Are We There Yet? The Long Road Toward Water Quality-Based Regulation Under the Clean Water Act. 27 *Environ. Law Report.*, 10391–10401.

Houck, O. (1997b) TMDLs: The Resurrection of Water Quality-Based Regulation Under the Clean Water Act. 27 *Environ. Law Report.*, 10329–10344.

Houck, O. (1998) TMDLs III: A New Framework for the Clean Water Act's Ambient Standards Program. 27 *Environ. Law Report.*, 10415–10443.

Houck, O. (1999) TMDLs IV: The Final Frontier. 27 *Environ. Law Report.*, 10469–10486.

Iron and Steel Manufacturing Point Source Category (2003) Code of Federal Regulations, Part 420, Title 40.

Michigan Department of Environmental Quality (2002) Rule Part 30: Water Quality Trading (effective November 22, 2002). Michigan Department of Environmental Quality: Surface Water Quality Division: Lansing, Michigan; http://www.state.mi.us/orr/emi/arcrules.asp?type=Numeric&id=1999&subId=1999%2D036+EQ&subCat=Admincode (accessed April 16, 2004).

National Academy of Public Administration (2000) *Transforming Environmental Protection for the 21st Century*; Washington, D.C.

National Research Council (2001) *Assessing the TMDL Approach to Water Quality Management*; National Academy Press: Washington, D.C.

Stacey, P., Connecticut Department of Environmental Protection, Hartford, Connecticut (2004) Personal communication.

U.S. Environmental Protection Agency (1986) Quality Criteria for Water; EPA-440/5-86-001; U.S. Environmental Protection Agency: Washington, D.C.

U.S. Environmental Protection Agency (1991a) Guidance for Water-Quality Based Decisions—The TMDL Process, Unpublished Guidance; U.S. Environmental Protection Agency: Washington, D.C.

U.S. Environmental Protection Agency (1991b) Technical Support Document for Water Quality-Based Toxics Control; EPA-505/2-90-001; National Service Center for Environmental Publications: Cincinnati, Ohio.

U.S. Environmental Protection Agency (1993) Guidance for 1994 Section 303(d) Lists, Unpublished Guidance; U.S. Environmental Protection Agency: Washington, D.C.

U.S. Environmental Protection Agency (1994) *Water Quality Standards Handbook*, 2nd ed.; EPA-823/B-94-005a; U.S. Environmental Protection Agency, Office of Water: Washington, D.C.

U.S. Environmental Protection Agency (1996) *Draft Framework for Watershed-Based Trading*; EPA-800/R-96-001; U.S. Environmental Protection Agency, Office of Water: Washington, D.C.

U.S. Environmental Protection Agency (1998a) *Fed. Regist.*, 63 (129), 36742, July 7.

U.S. Environmental Protection Agency (1998b) *National Strategy for the Development of Regional Nutrient Criteria*; EPA-822/F-98-002; U.S. Environmental Protection Agency, Office of Water: Washington, D.C.

U.S. Environmental Protection Agency (1998c) *Report of the Federal Advisory Committee on the Total Maximum Daily Load (TMDL) Program*; EPA-100/R-98-006; U.S. Environmental Protection Agency, Office of the Administrator: Washington, D.C.

U.S. Environmental Protection Agency (2001a) 2002 Integrated Water Quality Monitoring and Assessment Report Guidance. Unpublished Guidance. http://www.epa.gov/owow/tmdl/2002wqma.html (accessed April 16, 2004).

U.S. Environmental Protection Agency (2001b) *Chesapeake Bay Program Nutrient Trading Fundamental Principles and Guidelines*; EPA-903/B-01-001; Chesapeake Bay Program, Nutrient Trading Negotiation Team: Rockville, Maryland.

U.S. Environmental Protection Agency (2001c) National Nutrient Guidance Memo: Development and Adoption of Nutrient Criteria into Water Quality Standards. Unpublished Guidance.

http://www.epa.gov/waterscience/criteria/nutrient/guidance/index.html (accessed April 9, 2004).

U.S. Environmental Protection Agency (2002a) *Consolidated Assessment and Listing Methodology, Toward a Compendium of Best Practices*; U.S. Environmental Protection Agency: Washington, D.C.

U.S. Environmental Protection Agency (2002b) *Draft Strategy for Water Quality Standards and Criteria: Strengthening the Foundation of Programs to Protect and Restore the Nation's Waters*; EPA-823/R-02-001; U.S. Environmental Protection Agency: Washington, D.C.

U.S. Environmental Protection Agency (2002c) *The Twenty Needs Report: How Research Can Improve the TMDL Program*; EPA-841/B-02-002; U.S. Environmental Protection Agency: Washington, D.C.

U.S. Environmental Protection Agency (2003a) Guidance for 2004 Assessment, Listing and Reporting Requirements Pursuant to Sections 303(d) and 305(b) of the Clean Water Act; TMDL-01-03, Unpublished Guidance; U.S. Environmental Protection Agency: Washington, D.C.

U.S. Environmental Protection Agency (2003b) Water Quality Trading Policy, Unpublished Guidance. http://www.epa.gov/owow/watershed/trading/finalpolicy2003.html (accessed June 20, 2004).

Water Environment Federation (1997) *The Clean Water Act: 25th Anniversary Edition*; Water Environment Federation: Alexandria, Virginia.

Water Quality Planning and Management (2003) Code of Federal Regulations, Part 130, Title 40.

Water Body Use Designation (2002) Ohio Water Quality Standards, Administrative Code Chapter 3745-1; Ohio Environmental Protection Agency: Columbus, Ohio; http://www.epa.state.oh.us/dsw/rules/3745-1.html (accessed March 22, 2004).

CHAPTER FOUR

An Economic Framework for Evaluating Trading Opportunities

Introduction

This chapter presents an economic framework for assessing the advantages and disadvantages of trading (as a buyer, seller, or observer) and provides guidance on how to work through key considerations in identifying and evaluating options already available or that could be developed. The framework addresses incremental cost comparisons between trading and nontrading options and goes further to describe a broader assessment of the potential benefits and costs of trading options consistent with the type of capital planning and strategic decisionmaking processes that many utilities follow.

A sufficiently broad framework is needed because wastewater treatment plants (WWTPs), evaluating the desirability of water-quality trading and assessing the benefits and potential concerns about specific trading proposals, do so in the context of their organization's mission, goals and objectives, and specific responsibilities for various aspects of water-quality protection and watershed management. Wastewater treatment plants, publicly owned treatment works, and regional utilities responsible for wastewater treatment and stormwater management face a unique set of responsibilities as service providers subject to governmental oversight; as regulated entities subject to a set of complex and specific requirements; and as stewards of watershed resources for their customers and communities. Balancing these multiple responsibilities is challenging, even when no innovative approaches are being considered. Adding water-quality trading to the mix could make it more challenging, but it could also provide greater flexibility and additional alternatives for finding solutions that meet environmental, economic, and other objectives. As in other chapters, the term WWTP will be used to refer to the broad class of dischargers identified above.

An economic assessment of trading options typically comes into play when a WWTP's situation changes in a critical way that creates a need or opportunity for action. For example, WWTPs are frequently subject to lower limits, loading caps, or growth restrictions because of water-quality impairments in their receiving waters, imposed either through the normal permit renewal process or as the result of a total maximum daily load (TMDL) wasteload allocation. A WWTP also could be simply reevaluating its approach to meeting a current requirement for one or more pollutants.

In most such situations, if trading is an option, WWTPs have the following basic choices:

- ■ Upgrade technology to meet the new effluent limit or loading allocation;

- ■ Use current technology and create or purchase credits to offset loadings above the allocation;

- ■ Upgrade technology beyond that required to meet the allocation, and sell surplus reductions to other sources (or bank them); or

- ■ Appeal the allocation decision and delay action pending resolution, eventually choosing one of the above three options if some additional control or loading reductions are warranted.

In a trading framework, these choices represent three "market positions": nontrading observer, buyer, and seller. The best option may appear evident without detailed analysis—however, it is worth considering each position, at least at a preliminary level.

For example, it may be tempting to think of WWTPs primarily as buyers of credits. The stereotypical trading example is a WWTP buying credits from a farmer. In many watersheds, this may be an economically and environmentally effective solution. But in just as many watersheds, WWTPs may find the best trades among themselves, with other units of government, or by investing in stream, wetlands, or habitat restoration projects. Hence, WWTPs should keep an open mind about the prospects of buying credits, selling them, or not trading at all.

This chapter is organized around a set of key questions that form the basis for a series of steps in the evaluation framework. A WWTP can work through these steps in varying levels of detail, in a straight path, or iterative fashion, depending on whether the purpose is to get a preliminary sense of potential opportunities, refine good options, or develop a concrete trading proposal.

Chapter 5 presents a more trade-specific set of considerations focused on technical feasibility and selected aspects of executing and implementing trades, which includes additional discussion of economic and financial aspects of trading. Chapter 4 is intended to help develop the general analytical framework for the more detailed evaluations addressed in Chapter 5. Importantly, working through the issues identified in Chapter 4 before proceeding to Chapter 5 will help identify the type of economic and financial information needed to adequately evaluate specific trading proposals. In practice, after the first reading of these chapters and assimilation of the concepts introduced, the evaluation may be combined and conducted concurrently, as is also noted in Chapter 5.

The steps and corresponding questions presented in Chapter 4 are as follows:

(1) Estimate credit need. What is the scale and scope of the additional mass loading reductions (or other action) contemplated, and what are the economic and financial implications of that need?

(2) Identify trading options. What are the technically feasible alternatives to meet regulatory and other requirements, and what "market position" does each represent?

(3) Characterize trading options. How could a WWTP categorize and describe its feasible alternatives in terms of cost, effectiveness, and other key factors?

(4) Evaluate trading options: How do the alternatives compare to each other with respect to cost-effectiveness, market position, and other relevant decision criteria?

(5) Develop trading proposals. What are the most promising actions suggested by the evaluation, and how should the WWTP proceed?

Estimating Credit Need

To take most advantage of the framework described in this chapter, a WWTP must first have a general sense of the scale and scope of the new or changing pollutant control responsibility. This includes understanding the possible range of additional mass loading reductions that may be required, in absolute terms and as a percent of total loadings and any preexisting reduction target, and the level of financial investment likely necessary for compliance. For technical and financial reasons, the timeframe over which the new or existing target must be met also must be considered. The WWTP must consider whether it is

(1) Looking for a specific number of credits in the short-term;
(2) Making decisions about how many and what types of credits it would or could buy or create, and possibly sell over the medium term; or
(3) Developing a long-term strategy to optimize allocation of resources across nontrading and trading opportunities.

For example, WWTPs finding themselves in the first situation may need relatively few credits compared to their overall load, but may need to secure the credits for compliance within the current permit term (e.g., less than five years). This is likely to represent a minor to modest financial effect on the overall budget. Under these circumstances, a simple analysis may be sufficient, or time pressure or available data may not allow for a detailed benefit-cost comparison. The appropriate and feasible use of the framework described in this chapter will, therefore, probably focus on unit-based cost comparisons between in-plant options involving operations and maintenance (O&M) expenditures versus readily implementable projects involving some capital investment and O&M. It may not be possible, in these situations, to incorporate a significant number of other evaluation criteria, beyond cost-effectiveness and compliance certainty, in very much detail.

In contrast, many WWTPs find themselves in the second situation; for example, where trading is not needed for compliance with current permit limits, but is an option for the next permit; or perhaps in cases where TMDL wasteload allocations will result in lower permit limits. In these situations, the relative decrease in loading caps and the investment required for compliance will often be significant enough to justify, if not require, a more precise comparative analysis of the trading and nontrading options. The best use of the economic framework, in these cases, should involve cost-effectiveness comparisons at the dollar-per-kilogram (pound) level and a broader consideration of total costs and benefits over one or two permit cycles in the future. The options examined likely will include a range of O&M and capital investments at the facility compared to trading options that may involve other point sources and nonpoint sources. It should be possible, in these analyses, to incorporate decision criteria beyond cost-effectiveness, such as how well proposed trades support other WWTP mission objectives relating to watershed stewardship.

Wastewater treatment plants in the third situation above are probably involved in long-range strategic planning looking at a period covering more than two permit cycles (i.e., 10 to 20 years). They are essentially evaluating a water-quality management portfolio that includes all the different technological controls, process improvements, O&M options, and other alternatives a WWTP has at its disposal to meet permit limits and other watershed-related goals. This may mean examining what the water-quality management portfolio currently looks like and envisioning how it might be changed over time to achieve the greatest return on investment. Applying the evaluation framework presented here to this type of analysis is similar to the medium-term situation described above, except that the cumulative pollutant control responsibility and financial investments are much greater over the longer time frame. The number of alternative scenarios considered will probably be higher and the set of decision criteria should be richer. For example, there may be an interest and opportunity to consider long-term, non-water-quality benefits from flood control, habitat enhancement, and changes in agricultural practices that may be associated with certain credit-generating activities, in addition to cost-effectiveness.

The example situation presented in the following section is representative of a good number of WWTPs that have already been involved in or are contemplating trading programs. It will be used to illustrate results from following the framework presented in this chapter.

Example Wastewater Treatment Plant

The example WWTP currently has technology-based effluent limits expressed as a concentration limit, but it discharges to a river on the state's 303(d) list of impaired waters. This WWTP has 2½ years remaining on its current permit term. A TMDL will be completed for the WWTP's receiving water in time for the final wasteload allocation to be translated into new limits in the WWTP's next permit. The draft TMDL analysis has indicated that the WWTP may be responsible for a 35% reduction in its current average annual loadings, calculated over the last three years. The state regulatory agency is willing to sign a letter of agreement today, granting the WWTP the full five years of its next permit to comply with a new loading cap if the WWTP begins working toward that goal immediately, first with a preliminary economic analysis of compliance options, including trading, and then by implementing any feasible cost-effective reductions as soon as practicable. If experiences with trading in the short- to medium-term are successful, this WWTP would like to integrate trading options into its longer-range strategic and capital planning process. The WWTP is, therefore, interested in evaluating trading options using an economic framework that supports these analyses.

Identifying Trading Options

Identifying potential trading options consists of the following two components:

    (1)   Listing the possible sources of additional reduction or credit, in terms of source categories and generalized technical options; and

(2) Identifying whether the options put the WWTP in a market position of a buyer, seller, or nontrader.

Key considerations are outlined below, followed by Table 4.1, which summarizes the issues using the example introduced above.

Sources of information to identify technical options include the following:

(1) Internal planning and budget documents;
(2) Planning and budget documents from other public agencies;
(3) Interviews with key staff at public and private organizations;
(4) Publicly available regulatory documents and watershed management plans, especially including TMDL reports and implementation plans; and
(5) Information from local governments, non-profits, and state and federal agencies involved in land use management.

See also the discussions regarding trading options and potential partners in Chapter 5.

TABLE 4.1 Possible credit sources as they relate to market positions.

EXAMPLE PRELIMINARY LIST OF POLLUTANT REDUCTION OPTIONS	A. NONTRADER BASE CASE	B. BUYER CREDITS = > 0% OF COMPLIANCE STRATEGY	C. SELLER SURPLUS CONTROL, EXTRA CREDITS
Facility-based			
Pretreatment program changes Inflow volume reduction Collection system BMPs Optimize existing facilities	One or more of these certainly would be associated with not trading.	These would involve being a buyer if in-plant reductions alone were insufficient to meet the target.	These could involve being a seller if in-plant reductions were greater than needed to meet the target.

(continued on next page)

TABLE 4.1 Possible credit sources as they relate to market positions. *(Continued)*

EXAMPLE PRELIMINARY LIST OF POLLUTANT REDUCTION OPTIONS	A. NONTRADER BASE CASE	B. BUYER CREDITS = >0% OF COMPLIANCE STRATEGY	C. SELLER SURPLUS CONTROL, EXTRA CREDITS
Moderate capital and O&M Major capital upgrade			
Utility owned or directed			
Effluent reuse and recycling Flood control Stormwater BMPs Parks and recreation projects	May not be trading, depends on specific action or regulatory arrangement.	Buyer if it involves other governmental unit, other landowner, separate permit, etc.	Possible seller if A and/or B true, but more reductions than needed are created.
Point sources			
Other WWTPs Industrial Municipal stormwater Combined animal feeding operation	By definition, all involve trading.	By definition, all will involve buying.	These could involve selling if credits purchased were resold or transferred.
Nonpoint sources			
Ag-animal BMPs Ag-crop BMPs Erosion control measures Flow augmentation Reforestation Riparian corridor improvements Septic BMPs and conversions Urban stormwater BMPs Wetlands restoration	By definition, all involve trading (except perhaps if action fell under utility owned or directed and implemented in NPDES permit in such a way to not constitute trading).	By definition, all will involve buying (except perhaps for same reason as noted under nontrading case).	These could involve selling if purchased credits were not used for compliance and instead were resold or otherwise transferred.

POSSIBLE SOURCES OF REDUCTIONS OR CREDITS—TECHNICAL OPTIONS

Identifying the WWTP's potential trading options involves making a list of the technically feasible alternatives that are or may be available. Because trading initiatives are often an advancement to innovation, either at the WWTP or in the watershed, it may be valuable to initially brainstorm a list of options and identify everything that is technically feasible. The WWTP could then decide to prescreen the list before moving forward and eliminate those with a low probability of being viable (for example, not affordable under any circumstances or known to be politically impossible).

The list of technically feasible alternatives could be grouped into four major categories—facility-based, other utility owned or directed program, other point sources, and nonpoint sources—as described below and illustrated in Table 4.1. Each category could include low-, medium-, or high-cost alternatives, depending on the specific mix and type of facilities and operations.

■ Facility-based alternatives include minor to major changes, enhancements, and upgrades at the WWTP or at related facilities (collection system, pump stations, etc.). They represent investments in O&M and other noncapital expenses, such as staffing changes or instrumentation improvements, a combination of O&M and relatively low-cost capital expenses for enhancing selected unit processes, and include everything up to high-cost, capital-intensive facility upgrades.

■ Utility owned or directed alternatives include other pollutant reduction or other environmental improvement opportunities that some utilities have available to them within their organizational structure and/or geographic or infrastructure responsibilities that are not implemented at the wastewater treatment facility itself. Examples include effluent reuse, flood control, stormwater management, and source-water protection (for joint utilities). Technically, these may represent either a point or a nonpoint-source credit opportunity from a regulatory or physical definition.

Regardless of the technical status, it generally will be important to differentiate "self-generated" credit trading opportunities from those involving other parties, either as a separate utility category or as a subcategory under point or nonpoint sources.

■ Point-source alternatives include other WWTPs and industrial dischargers who could maintain loadings below their effluent limit, wasteload allocation, or voluntary loading cap. Following a strict regulatory definition, sources with municipal stormwater (MS4) and combined animal feeding operation permits also would be included in this category. However, if the WWTP wants to structure its list considering end-of-pipe control versus land-based control, some of these types of sources could be categorized as nonpoint sources, or in a separate National Pollutant Discharge Elimination System (NPDES) land-based category for evaluation purposes.

■ Nonpoint-source alternatives include the variety of in-stream, near-stream, and land-based best management practices (BMPs) and other improvement projects that may be available in the watershed, such as erosion control measures, flow augmentation, riparian restoration, wetlands restoration, agricultural BMPs for crop and animal operations, stormwater BMPs, and septic system improvements and conversions.

MARKET POSITIONS

The potentially feasible solutions from the four categories of technical options represent up to three distinct market positions.

(1) "No-trading" option(s). The WWTP can meet its objectives without trading. This option represents one or more base cases, against which the WWTP would evaluate the costs and benefits of the potentially viable trading options.

(2) Buying options. The WWTP purchases or otherwise arranges for credits from point, nonpoint, and/or third parties that offset loadings above a cap, where credit trading may represent a minor, moderate, or major portion of a utility's water quality management strategy (in terms of kilograms [pounds] and/or dollars represented).

(3) Selling options. The WWTP engages in a sales transaction or otherwise transfers surplus credits to point, nonpoint, and/or third parties (to be used to offset their loading above a cap), banked for later use (if allowed), or retired. These are credits the WWTP may have created by reducing loadings below its cap, by overbuying credits from someone else, or by purposefully implementing BMPs or watershed improvement project(s) that generate surplus credits.

POTENTIAL CREDIT SOURCES AS THEY RELATE TO THE WASTEWATER TREATMENT PLANT'S MARKET POSITION

It is important to consider the relationship between the various ways in which a WWTP might meet a new or continuing pollutant load reduction responsibility and its credit market position as a possible buyer, seller, or nontrader. As is discussed later, each market position carries with it a slightly different set of economic and financial considerations, and, with each position, the WWTP's ability to influence or directly affect various components of the credit costs also will vary. Additionally, depending on a WWTP's given situation and preferences, certain attributes or features of being a buyer, seller, or nontrader will be more or less important among the economic-related decision criteria. For these reasons, associating credit options with their corresponding market position(s) will help frame the analysis most appropriately. Table 4.1 presents a generalized example of one such set of relationships for the example WWTP introduced in the Example Wastewater Treatment Plant section.

The list in the left column of Table 4.1 includes the theoretically possible ways the example WWTP could create or arrange for pollutant-load reductions that would be creditable to its loading target. The example WWTP created this list by examining its own facility and the operations of other governmental entities in and near its service area and by reviewing a watershed management plan and the draft TMDL document prepared by the state regulatory agency. After creating the list, the WWTP made the notes presented in the corresponding market position columns to begin the process of considering the market position(s) most likely associated with each credit option.

Characterizing Trading Options

After a basic list of trading options has been identified, the alternatives must be more fully defined and described before they can be evaluated and compared. Characterizing trading options is a two-step process; the steps could occur sequentially or in parallel. The first step involves establishing the key considerations that will form the basis for the WWTP's decisionmaking. These considerations must be framed in terms of performance metrics and decision criteria that the WWTP can use to meaningfully assess the alternatives. Some decision factors may be directly translated into quantitative evaluation criteria, while other factors would be defined in narrative form. Once the criteria are established, the next step is to define or "grade" the options using the selected criteria; this step is described in the Evaluating Trading Options section of this chapter.

Clearly, the cost-effectiveness of various trading options is the predominant priority. However, cost-per-unit of reduction, by itself, rarely tells the whole story. First, a whole range of other factors influence the relative cost-effectiveness of any option. Second, a broad range of important considerations that cannot be monetized, and therefore can't be reflected in the cost-effectiveness estimate, generally exist.

When determining the applicable regulatory requirement, most WWTPs would begin evaluating the business case for trading by first asking whether they could save money by buying credits or if they could make money by selling them, compared to not trading at all. The inquiry may proceed along the following lines:

■ It would be very expensive to upgrade the treatment plant, and the elected officials and ratepayers think they already pay a lot. So, do less expensive alternatives exist?

■ The WWTP could upgrade to a state-of-the-art facility that would all but guarantee no future additional requirements; however, how is the cost justified, especially if this option produces a significant amount of "over-control" for some period of time?

■ The WWTP believes actions other than, or in addition to, reductions of pollutants at the outfall would be equally or more beneficial to water quality and watershed management objectives, and is willing to invest in these actions; is there a way to receive credit for doing so?

■ The WWTP management, board of directors, or customers are undecided, ambivalent, conflicted, or nowhere near consensus; are the economics compelling in one direction or the other?

In the simplest of economic frameworks, three components are considered for any option: costs; revenues or other things that offset, reduce, or avoid costs; and the environmental benefit, frequently referred to as effectiveness in shorthand. With these three components, cost-effectiveness can be measured for most options, to varying degrees of rigor and precision.

Most cost and some benefit elements for each trading option can be readily accommodated using a relatively straightforward comparative life-cycle cost analysis in a way that captures the most important considerations and facilitates the decisionmaking process. A life-cycle cost analysis is performed by projecting all costs associated with a particular trading option during a defined time period (revenues and monetizable benefits can be included in this analysis by treating them as negative costs). The time period selected for the evaluation will be driven by the time-frame during which the WWTP is considering trading options. Comparisons of trading options can thus be made by projecting the relevant costs during the forecast period, defining an annual cost stream, and calculating the net present value of projected costs for each trading option. From this analysis, cumulative, average, or period-specific estimates of cost-effectiveness can be calculated by dividing the cost by the relevant environmental performance metric, such as kilogram (pound) removed or controlled.

Because it is assumed most WWTPs (or their consultants) are well versed and experienced in calculating and comparing engineering and management options involving capital and non-capital costs over defined periods of time, such methods are not discussed in any further detail here. Additional examples of applications of these methodologies for trading analysis can be found in several Water Environment Research Foundation-sponsored studies, among other sources (Bacon and Pearson, 2002; Kieser, 2000; Moore et al., 2000; Paulson et al., 2000).

It may be necessary to enhance the simple framework described above to address nonmonetizable costs and benefits. This can be accomplished by using the approach to developing and applying decision criteria described in this chapter. Each WWTP that is evaluating trading options would need to go through a process to select the decision criteria that are most important in its situation. Different components that could be included in a more complex cost-effectiveness comparison will be more or less applicable, depending on whether the WWTP is evaluating specific credit purchases, medium-term options, or longer-term strategies.

Potentially applicable considerations that could form the basis for decision criteria are discussed in more detail below under headings relating to costs, benefits, and effectiveness. In addition, Table 4.2 illustrates, in a summary fashion, how these considerations could be used to describe selected trading and nontrading options.

COST AND COST-RELATED FACTORS

Cost-effectiveness depends on a number of things that affect the cost side of the equation, including but not limited to the following:

■ Financing options;

■ Funding sources;

TABLE 4.2   Characterizing trading options.

EXAMPLE COST AND BENEFIT CONSIDERATIONS	EXAMPLE TRADING OPTION CATEGORIES (SEE ALSO TABLE 4.1)			
	Facility-based	Utility owned or directed	Other point sources	Nonpoint sources
Costs				
Capital Operation and maintenance Funding Financing Project delivery Procurement Schedule Staffing and labor Trading ratios Transaction costs Liability costs Deferring requirements	Capital, O&M, and implementation costs will range across the options. No trading transaction costs. Liability does not change from status quo (except perhaps if selling credits) costs.	A range of costs. Assumed off-site of WWTP, so ratios, transaction costs, and different liability schemes are likely. If land-based, financing and delivery mechanisms may be different than at WWTP.	Same issues as at WWTP, but may not see as buyer if paying $/kg(lb). Ratios and transaction costs are likely. Liability depends on arrangement between traders and how treated in NPDES . permits	Costs likely to range across options. Some considerations may not be applicable if paying per credit. Ratios and transaction costs are likely. Liability costs would depend on specific arrangement.
Monetized benefits				
Water quality credit sales Other credit sales Ancillary benefits	Sales possible. Ancillary benefits are limited to none.	Sales possible. Other benefits are project-specific.	Only resales. No additional benefits.	Sales possible. High opportunity for other benefits.
Unmonetized benefits				
Utility watershed stewardship Safe Drinking Water Act (1974),	Can support local, regional economic development,	Can support stewardship beyond plant footprint.	Similar to facility-based. Could further environmental	Strongest benefits related to watershed

(continued on next page)

TABLE 4.2 Characterizing trading options. (*Continued*)

EXAMPLE COST AND BENEFIT CONSIDERATIONS	EXAMPLE TRADING OPTION CATEGORIES (SEE ALSO TABLE 4.1)			
	Facility-based	Utility owned or directed	Other point sources	Nonpoint sources
Endangered Species Act (1973), etc. Economic development Environmental justice Public education Adaptive management Scientific gains	and environmental justice goals. Can further technology innovation at plants.	Can provide education and adaptive management opportunities.	justice, education, and adaptive management if location is better than at WWTP.	stewardship drinking water protection, and habitat enhancement.
Effectiveness				
Capacity for scalability and divisibility Reduction profile over time Variability	Scalability and capacity depends on plant age and configuration.	Wide range possible, depending on site and project type.	Generally the same as for WWTP, but may provide different options.	Capacity depends on non-point-source trading baselines. Other items in a range.

- ■ Implementation schedules;
- ■ Project delivery options;
- ■ Procurement alternatives;
- ■ Staffing effects;
- ■ Market fluctuations in variable-cost items (e.g., electricity, chemicals);

■ Variability in process performance;

■ Liability;

■ Risk; and

■ Transaction costs associated with implementing trading options.

It is also necessary to consider the extent that these costs are above or different from those incurred without trading, as opposed to being incurred with or without trading.

Base Costs

A comparison of trading alternatives—as a buyer and/or seller—to the no-trading option must include a full cost evaluation that considers capital, interest, O&M and labor, and transaction costs and risk insurance. The cost evaluation should be done equally and completely for both the trading and nontrading alternatives. For facility-based options, this should include any additional costs that may be associated with effects on other aspects of plant operations not directly targeted, such as changes in biosolids production and disposition. If an initial screening comparing unit treatment costs (i.e., cost per unit of mass per unit of time, such as $/kg[lb]/yr) is favorable, then a full cost analysis must be performed to truly determine whether trading options are attractive.

Funding Sources

The source of funds used for trading options could influence total costs based on whether a particular option is eligible for a specific funding source and if there are any restrictions. Typically available funding options for facility-based options include rates and charges, effect or system development charges, connection fees, grants, and developer contributions. Where grants are used, for example, they would directly lower the effective cost of an option to a utility.

In these situations, grant-funding may reduce the number of credits generated or transferable based on applicable policies or rules. For non-facility based options, a WWTP may fund credit creation or credit purchases from third parties with operating funds and/or debt financing, depending on the amount of money involved relative to its operating and capital budgets.

Financing Options

A short list of alternative sources of financing for planned capital improvements may include revenue bonds, general obligation bonds, state revolving fund (SRF) loans, and assessment bonds. Compared to "pay-as-you-go" financing, debt funding will generally involve interest-related costs but may enable quicker implementation. The different options may carry different interest rates. More important are the eligible activities that may be funded with each financing option. For example, SRF loans may offer a lower interest rate than revenue bonds, but it may not be possible to use the loan proceeds to purchase credits from another source or institution.

Project Delivery Options

A range of project delivery options may be available to the WWTP or its trading partners for point-source or selected nonpoint-source trading options that could influence funding sources, the cost of capital, or the implementation schedule. Examples include traditional design–bid–build approaches, construction management, design–construction management–general contractor, design–build, and fully integrated design–build–operate options. Increasingly, utilities and other organizations responsible for large watershed management programs are relying on alternative project delivery methods to provide substantial cost savings, reduce and transfer risks, and predetermine utilities' financial obligations.

Procurement Process

The contract instrument used for securing delivery of the project-generating credits may govern throughout project development. Time is one consideration. For example, some facility-based options may be more quickly procured than others. This affects both the cost of the actual procurement process and, potentially, the net present value of a control alternative. Purchasing credits from other sources may involve contractual mechanisms with which a utility is familiar, or it may require new approaches. This may be especially true where a WWTP engages a second party (e.g., the landowner) or third party (e.g., a contractor or nonprofit) to develop and implement a non-point-source BMP. Contractual approaches to implement trades are discussed in more detail in Chapter 5.

Implementation Schedules

Attention must be given to how the various trading options would affect the schedule for achieving the pollutant loading reductions or other improvements. There may be compliance deadlines or financial or other penalties for not reaching milestones. Additionally, the time-value of money would affect the net-present-value calculations. For example, some credits may be immediately available on a trading market, but at a premium cost, compared with a facility-based option that would cost less but take longer to implement. Alternatively, some facility-based options would produce immediate reductions in pollutant loadings at the outfall, as compared to an option like a riparian restoration project, which would take longer to become established and generate reductions.

Staffing Effects

Would the effect of alternative control and trading options be positive, negative, or neutral in terms of man-hours or required skills?

For example, non-capital-intensive, facility-based options may require additional training, reallocation of staffing (perhaps an additional full-time equivalent on staff), and/or some consulting or contract assistance. Major upgrades could have a significant effect on staffing regarding O&M. By comparison, purchasing credits from point and/or nonpoint sources likely would involve no additional facilities staff, but could require additional time from management, legal counsel, or consulting support to develop and implement the trading option.

Trading Ratios

Both unit-cost screening and total-cost analysis must account for any trading ratios that increase (or, less often, decrease) the number of kilograms (pounds) of pollutant reduction for each kilogram (pound) of credit the WWTP receives. Trading ratios and units are discussed in more detail in Chapter 5.

Transaction Costs

These include costs associated with implementing trading options above or different from those incurred without trading. The base case(s) would establish the baseline level of transaction costs. Additional costs may include development of special permits or agreements to implement trades, credit brokers or consultants fees, insurance or performance bonds, and additional documentation or reporting. Trading options could have lower transaction costs compared with the base case. For example, when comparing a facility upgrade to purchasing credits, the credit purchases likely would not involve the same level of expenditures and in-kind resources required for NPDES and construction permits.

A variety of means may be available to help minimize transaction costs. The WWTP should consider the following questions when estimating transaction costs that may be associated with different trading and nontrading options:

- How much time and effort would it take to find credits and execute the deal?

- How much paperwork would be involved?

- Could the WWTP go to a Web-based marketplace, identify trading partners, and buy credits?

- How often and in what way would the WWTP need to document trades—every trade, every quarter, or annually?

- Would electronic filing, recording, and reporting mechanisms be available to facilitate trading? Would every trade have to be reflected in the WWTP's permit, or would the WWTP have a variable limit or other clause that allows some or all trading to occur without permit revisions?

Liability and Risk

Liability and risk costs could be considered a subset of transaction costs or separated for more specific consideration. If the WWTP is a credit buyer, the considerations include the following:

- The extent of its liability, in the event that the credits purchased are later determined to be technically invalid;

- The project that generated them fails to properly operate or is not properly maintained;

- Contractual or legal mechanisms that would be available to the WWTP and whether the trading program has procedures in place to facilitate resolution; and

- How much time the WWTP would have before a violation or noncompliance is invoked.

As a credit seller, the considerations are similar; however, because point-source reductions are more readily measured and documentable within the existing NPDES framework, there has been less concern over these issues. Trading agreements between the WWTP and its credit buyer(s) should address credit life and the extent and duration of the seller's obligation to provide credits. For credit sellers, mechanisms to mitigate liability and risk include credit certification, contract provisions covering "acts of God" and other unusual or uncontrollable events, and insurance or performance bonds to help cover any additional expenses associated with these situations.

These liability issues associated with buying credits, the ways in which the WWTP can avoid them or minimize their effect, and risks that may be associated with WWTPs selling credits are discussed in greater detail in Chapter 5.

Cost of Deferring Requirements

The cost of deferring requirements may include additional study, administrative appeals, and litigation to determine whether the wasteload allocations and effluent limits were properly and fairly set. The WWTP would have to weigh the cost of this process against how much more or less it would end up spending to meet the final requirements. This approach is not necessarily recommended; however, it is recognized as a potential course of action available to most WWTPs. For example, if a WWTP feels its wasteload allocation is not technically valid, this approach may be warranted, and associated costs should be included as appropriate in the WWTP's trading evaluation.

ON THE PLUS SIDE—REVENUE OPPORTUNITIES AND OTHER MONETIZABLE BENEFITS

Revenue from Water-Quality Credit Sales

If the WWTP elects a trading option that involves creating surplus reductions that are creditable under a local trading program or state-permitting program and sold in market-based situations, it could count the income from the projected credit sales against the cost of the trading option. There are other situations in existence or being contemplated where WWTPs create surplus credits and transfer them to another governmental unit or other WWTP, under an arrangement that does not involve a direct transfer payment. If options include such an arrangement, the WWTP would have to characterize the benefit using nonmonetary or qualitative descriptions.

Revenue from Multicredit Sales

Under some circumstances, it may be possible to implement a non-point-source project that provides benefits beyond pollutant load reductions, including, for example, wetland restoration, habitat enhancement, and/or carbon sequestration. If the demand for these types of benefits exists in its area and there is a preexisting or creatable mechanism to market and sell those credits, the WWTP may be able to sell other types of credits. For example, if its riparian restoration project involved habitat improvement and tree planting, the WWTP might be able to sell habitat credits to a conservation bank and carbon credits to an electric utility (Bacon et al., 2003).

Monetized Value of Ancillary Benefits

Some trading options, including those involving nonpoint sources and other governmental programs, such as parks and recreation and fish and wildlife services, often involve benefits beyond loading reductions that could be monetized. These may or may not directly reduce the cost of a trading option to the utility. For example, stormwater BMPs implemented on county parkland or riparian restoration in a fishing area may lead to negotiating a transfer payment to the utility that is funded by an increase in user fees.

BENEFITS NOT ALWAYS EASILY MONETIZED BUT GENERALLY QUANTIFIABLE

Beyond cost-effectiveness, most utilities would want to evaluate trading options within the broader context of their overall mission—across environmental, social, political, and community responsibilities and objectives. Some of these are monetizeable, some are quantifiable in other ways, and some could be characterized using nonquantitative measures of cost or benefit. Examples of these broader considerations include the following:

- Contributions to a utility's or state's watershed stewardship programs;

- Support for other local, state, and federal water resource efforts covered under the Safe Drinking Water Act (1974), the Endangered Species Act (1973), wetlands restoration, and other programs;

- The speed with which pollutant reductions would produce water quality improvements;

- Environmental justice;

- Contribution to public education;

- Advancement of adaptive management principles; and

- Support for economic development and sustainable growth, including facility expansion and new facilities.

The last consideration has been receiving greater attention over the past several years, as water-quality concerns in some areas have resulted in local government restrictions on growth that have, in part, been implemented through NPDES permit restrictions, either with water-quality-based effluent limits or TMDL wasteload allocations. Increases in demand for wastewater treatment services (and, therefore, flow and loadings) stems directly from changes in residential and commercial development and, in some communities, large-scale development projects, including major subdivisions and major industrial and commercial sitings or expansions.

A WWTP serving such communities could consider trading (as a buyer or seller) as part of its strategy to help accommodate growth in a manner consistent with achieving watershed resource goals. Economic and financial attributes of various trading options could be more or less favorable in growth-pressured areas, and the WWTP will have to carefully examine how they play out in its watershed. For example, if growth patterns are well-understood and well-tracked, it is possible that a demand for credits to offset growth of existing communities and businesses, or to enable new development, could create incentives for partnerships in credit creation and facilitate efficient market transactions, leading to very cost-effective trading options. However, in situations where growth is less predictable, or occurs in spurts, fewer trading partners and trades could mean relatively higher transaction costs and credit prices.

ENVIRONMENTAL EFFECTIVENESS: THE DIVISOR IN THE COST-EFFECTIVENESS EQUATION

Capacity for Reductions

For each defined trading option, the WWTP should estimate the pollutant loading reduction or quantified environmental improvement that would result. This provides an assessment of the potential supply of different types of credits, on an individual source basis and within the watershed. These results must be specified in appropriate performance measures for the relevant period of time. Example results include 45 359 kg (100 000 lb) of phosphorus per year; 1000 kg (2205 lb) of nitrogen a month; and 100 kcal per day. The WWTP will have to identify the trading baseline of any potential trading partners. This is the level of pollutant control responsibility that a source must first meet, before generating surplus pollutant reductions (or other action) that may be creditable. Trading baselines were discussed in Chapter 3 and are also discussed in Chapter 5.

Scalability and Divisibility

For each component of a trading option or control alternative, what are the project sizes that are feasible or available? Are there a minimum number of credits that would be generated; is there a maximum number? For nontrading options or situations where the WWTP could sell credits, key questions focus on how scalable plant improvements or upgrades are. For any given "benchmark" BMP, such as a 3 ˘ 15-m (10 ˘ 50-ft) riparian buffer, how many of those could conceivably be implemented that are not already required to meet someone else's TMDL load allocation or other requirement? The flexibility of sizing credit blocks would vary across the trading options.

Reduction Profile over Time

What are the near-term estimates of water-quality credits generated by a particular option, and how is that level expected to change over time? For example, if the WWTP generates credits or buys them from another point source, the number of credits available may decrease over time if the flows increase and the treatment level (mg/L) remains the same. If the WWTP is implementing its own nonpoint-source project, then it may expect an increase in credits generated after initial installation, as the BMP becomes established, followed by a plateau level of effectiveness, and, later, a declining level of reductions even with adequate O&M. Where water-quality credits are available from a watershed market, clearinghouse, or third party that is developing a significant number of credit-generating projects, any actual variability of a project's performance over time may be mitigated or accounted for in the packaging of credits negotiated or offered for sale.

Variability

This characteristic of trading options is related to the reduction or credit generation or availability profile over time. For example, credit production from point sources could decrease over time if the selling point source's flow is increasing, but treatment levels (mg/L) stay constant. In cases where credits come from nonpoint-sources or land-based activities, credit generation could easily vary from year to year and across seasons, where loadings and BMP efficiencies are weather-dependent. The WWTP could estimate or negotiate an expected credit profile. Describing the potential variability of those values could involve a formal statistical analysis or a professional judgment about the extent to which the projected credit values might be lower or higher. Having some sense of the variability would allow the WWTP to develop credit profiles that represent more than point estimates to better support sensitivity analysis and decisionmaking.

EVALUATING THE COST AND BENEFITS OF TRADING OPTIONS

Characterizing trading options in terms of the relevant factors may be a one-time or iterative process for the WWTP, as it screens potentially feasible options, screens out infeasible and low-benefit ones, and progresses into a detailed assessment of the most promising options to add to its water-quality management portfolio. As suggested for the process of identifying options, at first, the WWTP may want to evaluate its active set of options, using all the possible decision factors. As it does so, the WWTP may find that some factors are more important than others in representing its primary goals and objectives. Additionally, in the longer list of potential concerns discussed above, some issues are interrelated, so it may be possible to consolidate some factors to streamline and focus the characterization of options.

Table 4.2 is a summary of what the WWTP might find when it evaluates its general categories of trading options and considerations the factors discussed above. The list in the left column includes the cost and benefit considerations discussed in the Characterizing Trading Options section of this chapter. The example WWTP conducted an initial screening of the issues and compared the four trading option categories examined in Table 4.1 across the general categories of considerations. The summary is simplified for purposes of presentation. A more detailed screening could be used to help develop more specific trading options and evaluation criteria, as discussed in the Evaluating Trading Options section of this chapter. Table 4.2 lays the groundwork for translating critical concerns and key attributes into decision criteria in the next section. In practice, Table 4.2 might represent a screening level examination or first cut, as a more detailed assessment will be necessary in most circumstances.

Evaluating Trading Options

After completing its identification and characterization of potential trading options, the WWTP would have determined whether it has a few, several, many, or perhaps no viable trading options. The evaluation framework selected should match the options list in terms of its complexity and level of data and effort required. A relatively simple framework may suffice if the WWTP has only a few options and a few key decision factors. Alternatively, a more structured and systematic evaluation framework would be warranted if, for example, one or more of the following were true: several options are viable; different time periods are considered; more than one kind of credit trading is involved; and the selected set of decision criteria are complex (some are competing, some are not readily quantifiable, etc.)

The evaluation framework described below provides a structured prioritization decision process for consideration of trading opportunities. By drawing on multiattribute utility analysis (MUA) approaches, it offers a systematic process specifically designed to address the problem of assessing the relative benefits, programs, or projects for organizations with multiple (and often competing) objectives. Using this type of process would help ensure that the selected trading option(s) is consistent with organizational objectives, that selections are based on well-defined measures of project performance, and that interrelationships between the utility's activities and those of other watershed stakeholders are recognized and accommodated. A significant benefit of the approach, especially applied within a public utility context, is the ability to incorporate consideration of both nonmonetary and monetary effects to a single analytical framework (Rothstein and Kiyosaki, 2003).

The WWTP can set up and apply this evaluation framework in the following basic steps:

(1) Define the specific **trading options** to be evaluated;
(2) Develop **decision criteria** from the list of considerations used to characterize the trading options; and

(3) **Rank the options** to compare how well they meet the WWTPs priority concerns.

These steps are described in more detail below. The discussion will address how it is possible to tailor this framework for a few options with simple criteria or for a larger number of options and more complex decision factors.

DEFINING THE TRADING OPTIONS TO BE EVALUATED

Before beginning the evaluation, it will be necessary for the WWTP to define its potentially feasible trading options in sufficient detail to assess how well each meets key decision criteria. The characterization of general trading options along key considerations, as described in the Identifying Trading Options section of this chapter, could be used as guide. In a more iterative process, the WWTP also could use a strawman set of criteria to refine a first-cut list of trading options into options suitably defined for evaluation with the final criteria. Whether the process is linear or iterative will depend, in part, on the amount of information initially available about trading opportunities and it will depend on the WWTP's (and other stakeholders') prior knowledge of and experience with trading. As a result, the WWTP could switch steps 1 and 2 as ordered above, or could iterate between them before proceeding to step 3.

The WWTP could identify discrete, mutually exclusive options, or it could identify several trading options that essentially represent a continuum of water-quality management portfolios that rely on trading to a lesser or greater extent. The choice would depend, in part, on whether the WWTP is in an early stage of its trading analysis, or if it has arrived at the final decision on whether to trade. Also, the WWTP should not be afraid to revise its options list upon reaching step 3 after an initial evaluation; the process of developing and applying evaluation criteria could help refine and optimize options.

Table 4.3 presents an illustrative set of trading options for the example WWTP. For purposes of illustrating the evaluation framework, refer back to the example WWTP introduced in the Estimating

TABLE 4.3 Base case and trading options defined.

OPTION	DEFINITION
A. Upgrade incrementally	This would involve phasing or staging upgrades such that loading targets would always be achieved with no need to buy credits. This level of upgrade would not provide any credit sales opportunity. This is the definitive "no trading" market position.
B. Upgrade to the limit of technology	This would involve significant capital investment in the plant, provide compliance with permit limits for the foreseeable future, and, at least for some period of time, provide the WWTP with surplus credits it could potentially sell to other sources. This would represent a "no trading" market position, with the possibility of also being a seller.
C. Facility optimization with low-to-moderate credit purchases	This would involve investing in the WWTP over time to improve treatment operations, but only to such a level that some amount of credit purchases alsowould be necessary to meet permit limits. During different periods, the optimization may involve only O&M improvements and/or low- to medium-cost capital investments. This represents a moderate buyer position, where credit purchases would represent less than one-half of the WWTP's load reduction responsibility.

(continued on next page)

Credit Need section of this chapter, and discussed again in the Identifying Trading Options and Characterizing Trading Options sections. Recall that this WWTP is facing a likely requirement to reduce loadings of a pollutant by at least 35% during the next seven years, and may have to do more. Based on a screening level identification and characterization of trading options, the WWTP has selected the five options listed in the table for further evaluation.

SELECTING THE DECISION CRITERIA

An example of a relatively simple set of decision criteria would be to select the least-cost solution that meets the permit limit. Another would be to evaluate the costs and benefits of viable trading

TABLE 4.3   Base case and trading options defined. *(Continued)*

OPTION	DEFINITION
D. Facility investment with significant point-source credit purchases	This would involve some investment in the plant over time, but only at the level necessary to maintain compliance in other areas and with pollutants other than the one(s) considered for trading. The WWTP would need to secure well over one-half of its required load reduction through credit purchases. In this buyer market position, credit purchases could represent as much as 35% of the WWTP's baseline load (in this example).
E. Facility investment with significant nonpoint-	This would involve a strategy of investing in the plant, similar to option D, except that the source credit purchases majority of the credits purchased would come from nonpoint sources. The example WWTP has the luxury of considering both point- and nonpoint-source credits, and has chosen to do so to better evaluate some of the additional benefits that non-point-source projects can provide to the watershed. As in option D, this represents a buyer market position, where credit purchases could represent as much as 35% of the example WWTP's baseline load.

options purely in terms of readily monetizeable factors and select the one(s) that had the highest cost-effectiveness or highest benefit-to-cost ratio.

However, because trading by definition involves partnerships, evaluating trading options generally involves decision factors beyond cost-effectiveness, as typically measured. This will be especially true when evaluating trading options over a medium- and long-term period. From the characterization of its trading options, most WWTPs would probably identify at least one important concern that is not strictly economic or financial. Utilities with active watershed management programs and strong stakeholder involvement programs would likely find several such nonmonetized critical concerns. In these circumstances, a more detailed set of decision criteria is needed.

A good set of decision criteria reflect the WWTP's critical objectives and priorities. In most circumstances, the list of individual considerations used to characterize the potential trading options would be too long and unwieldy to use to evaluate the options. However, within this list, the WWTP should find a smaller or consolidated set of considerations that reflect its organizational values, core mission, and water-quality-related goals and objectives. Distilling these key decision factors from the larger list of considerations provides the basis for setting formal evaluation criteria.

A well-constructed set of decision criteria has several characteristics (Rogers et al., 1997).

- ◼ Fundamental criteria reflect the WWTP's mission and what is important for it to accomplish.

- ◼ Comprehensive criteria cover all of the major concerns and policy issues that stakeholders consider to be of utmost importance, include all important dimensions of the consequences of the trading options.

- ◼ Relevant criteria are specifically influenced or affected by the feasible trading options.

- ◼ Well-defined criteria are articulated to facilitate communication with decisionmakers and stakeholders.

- ◼ Independent criteria do not address the same or overlapping aspects of WWTP performance, thereby avoiding redundancy or double-counting.

- ◼ Measurable criteria could be described using performance measures (monetary or nonmonetary), and are preferably quantifiable.

- ◼ Concise criteria. By following the other suggestions, the WWTP should end up with approximately six or seven fundamental decision criteria, with any subordinate criteria organized accordingly (in most cases, 10 should be sufficient; more than that is likely to mean some are redundant or not independent of another criterion).

For any evaluation of trading options, cost and cost-effectiveness will be key criteria in most situations. There are two ways to deal with these factors in the set of decision criteria:

(1) Cost-inclusive analysis. Cost or cost-effectiveness is included as a criterion in the decision model and evaluated directly alongside the other criteria.

(2) Value or cost analysis. Cost and cost-effectiveness are not included as criteria in the decision model. Instead, the set of decision criteria is constructed to define benefits associated with noncost factors. Cost and cost-effectiveness are determined separately and are then considered against the results of the criteria-based evaluation. Some believe this approach more directly weighs trade-offs between cost and other values.

Effective evaluation criteria will reflect the WWTP's fundamental objectives and help align decisions about trading options with its strategic direction. The criteria must be constructed to facilitate rating of trading options so that choices can be assessed as yielding greater or less benefit than the nontrading option. If better alternatives exist, the criteria also must be defined in a way that helps the WWTP select one trading option, or one option package, over others.

To illustrate these principles, a set of decision criteria for the example WWTP is listed in Table 4.4. This set of decision criteria was developed for the example WWTP using the approach presented above. The criteria attempt to capture the WWTP's water-quality, management-related objectives and how it measures its performance in achieving those objectives. In this case, the return on the WWTP's investment in a trading or nontrading option designed to meet its permit limits is, in fact, a function of multiple goals reflecting a broad diversity of community values. This example utility is challenged with not only efficient, cost-effective, reliable-service delivery, but also with protection of public health and environmental resources, community service, and public education. This set of decision criteria, therefore, reflects the example WWTP's intent to balance watershed stewardship with economic development and translates the WWTP's values into measurable benefits. Note that permit compliance, while obviously the most important concern, is not among the criteria. In this example, the WWTP has set compliance as a threshold criterion—no option would be evaluated that does not pass a compliance test. Alternatively, permit compliance (or, more accurately, the risk of failing to comply) could be included among the decision criteria and evaluated with the others. Each WWTP will need to determine which approach best serves its circumstance and preference.

The presentation of the example criteria in Table 4.4 is purposely listed in alphabetical order to give no indication as to the relative importance of one criterion versus another. In this example and in practice, equality may or may not exist. By taking another step and defining criteria weights, the WWTP could quantitatively express the relative value it places on each criterion. The weights would reflect how the WWTP would trade off one objective for another, when it evaluates trading options with competing objectives and multiple decision attributes (Rogers et al., 1997).

One straightforward way to establish criteria weights is to use a scale from 0 to 100. The criteria can be listed in order of priority and importance, and a weight of 100 can be assigned to the most important criterion. Then, the second most important criterion can be weighted relative to the first, the third relative to the second, and so on. Once the WWTP has assigned weights to each criterion, the overall ratings should be reviewed for consistency and validity. Lower-ranked criteria should be compared to higher-ranked ones to see if the relative weightings make sense, and the weights should be refined if necessary (Rogers et al., 1997).

Table 4.5 illustrates how the example WWTP might weight the criteria presented in Table 4.4 to reflect its values, concerns, and priorities. In practice, these could vary considerably based on a WWTP's specific situation.

Establishing numerical weights is useful when the WWTP is considering a greater number of options than is easily evaluated with narrative notations about the relative importance of criteria. Weights also

TABLE 4.4 Selecting decision criteria.

EXAMPLE CRITERIA (IN ALPHABETICAL ORDER)	EXAMPLE NARRATIVE DEFINITION REFLECTING THE WASTEWATER TREATMENT PLANT'S GOALS AND OBJECTIVES
Credit or reduction certainty	Level of confidence that the supply and availability of credits will be sufficient, or that selected technology, BMP, or other project will produce the estimated level of pollutant reductions at the plant.
Cost-effectiveness	Return on investment in terms of kilograms (pounds) of reduction per dollar spent; get the most "bang for the buck" during a given time period.
Liability and risk	Minimize or manage any additional actual or potential responsibilities associated with trading.
Overcompliance	Spend resources as efficiently as possible to minimize unnecessary or poorly-timed expenditures.
Pace of water-quality improvements	Hasten positive effects on water quality ahead of the TMDL and during implementation of the wasteload allocation.
Partnership opportunities	Partner with other watershed stakeholders to build institutional relationships and leverage informational and financial resources.
Watershed stewardship	Advance watershed management goals beyond effluent reductions, including flood control, habitat restoration, and recreational enhancements.

facilitate a quantitative ranking of the options being considered (as shown in Table 4.6). Other approaches are certainly valid and may be more appropriate in some situations (i.e., identifying decision criteria as being of high, medium, or low importance, or listing them in relative order of priority without assigning a weight). It also is possible to compare options strictly on the basis of monetizeable costs and benefits, such as unit cost-effectiveness or total pollutant load reduction

TABLE 4.5   Weighting decision criteria.

EXAMPLE CRITERIA (IN ORDER OF IMPORTANCE)	EXAMPLE WEIGHT	EXAMPLE RATIONALE
Credit or reduction certainty	100	The WWTP considers this the most important consideration for any trading or non-trading option.
Pace of pollutant-loading reductions	90	Because the state regulatory agency has offered a longer compliance schedule for early action, the WWTP is interested in options that provide reductions in a relatively short timeframe.
Cost-effectiveness	80	Not surprisingly, cost-effectiveness is a key priority, but the WWTP is willing to sacrifice some cost-effectiveness for certainty and quickness.
Overcompliance	80	Equally important as cost-effectiveness is not overcontrolling beyond the WWTP's reasonable loading reduction targets, at least not without receiving credit, in the form of credit sales or perhaps other regulatory flexibility.
Watershed stewardship	75	The WWTP is very interested in projects that would provide benefits beyond mass loading reductions.
Liability and risk	70	Minimizing extra liability that may be associated with trading is a concern, but one that the WWTP assumes can be managed through permit language and other agreements, so is less of a concern.

(continued on next page)

TABLE 4.5 Weighting decision criteria. *(Continued)*

EXAMPLE CRITERIA (IN ORDER OF IMPORTANCE)	EXAMPLE WEIGHT	EXAMPLE RATIONALE
Partnership opportunities	70	The WWTP is very interested in partnering with other stakeholders, but this does not take priority over the other considerations.

TABLE 4.6 Scoring the trading and nontrading options using decision criteria.

	REPRESENTED MARKET POSITIONS	NON-TRADING BASE CASE	NON-TRADING BUT POSSIBLE SELLER	CREDIT BUYING POSITIONS (SELLING OPTION FROM RESALES)		
Decision criteria	1 Weight	2 Upgrade in stages	3 Upgrade to the limit of technology	4 Optimize wastewater treatment plant, plus buy some credits	5 Facility investment with significant point-source credits	6 Facility investment with significant nonpoint-source credits
Credit or reduction certainty	100	10	10	8	7	6
Pace of pollutant-loading reductions	90	8	5	10	7	7
Cost-effectiveness	80	7	5	10	6	7
Overcompliance	80	7	5	7	8	10
Watershed stewardship	75	1	1	7	1	10
Liability and risk	70	10	10	9	8	5
Partnership opportunities	70	1	1	5	7	10

"return" on an investment, and list the unmonetized benefits in narrative form or as a series of pluses to be considered. As mentioned earlier, these simpler methods are not discussed in detail in this chapter, as it is assumed most WWTPs will be familiar with them.

APPLYING THE DECISION CRITERIA AND EVALUATING THE OPTIONS

Regardless of whether the criteria are assumed to be equally important, are numerically weighted, or their relative importance is indicated in some other way, it will be necessary to score, grade, or otherwise assess the trading and nontrading options along the selected decision criteria. As with the weighting exercise, quantitative or narrative scores could be given, and the WWTP will need to select the approach with which it (and its stakeholders) is most comfortable. One advantage of using a quantitative approach is that it facilitates tracking the evaluation process and comparing the multiple options. The WWTP has the option of using the quantitative scores and resulting priority rankings as a way of testing or cross-checking a less mathematical evaluation approach, or it could use the results as the basis for making decisions and a record to document the process for stakeholders.

If the WWTP elects to establish a quantitative scale for each criterion to score each trading option, two types of scales (also sometimes referred to as performance measures) could be used: natural or constructed scales.

(1) Natural scales. The performance or benefit could be expressed in common, generally quantifiable units, such as dollars, net present value, number of kilograms (pounds) of pollutant reduction, or acres of trees planted. These are used when direct measures or data are available.

(2) Constructed scales. The performance or benefit is not easily quantified using common metrics, and, instead, the scale reflects a narrative description of performance, with reference to specific criteria, and provides precise, unambiguous definitions of project performance. These can be used to

quantify expert opinions about performance in the absence of direct measurement.

Each evaluation criterion will likely have one or more clearly appropriate or feasible unit of scoring performance. For example, cost might be measured by dollars expended or percent savings over a given period of time. However, if the WWTP wishes to compare alternatives across the different evaluation criteria, it would need to convert the different scales into a common scale. In the MUA-based framework cited in the beginning of the Evaluating Trading Options section, a scale of 0 to 10 is typically selected. For each criterion, different levels of performance can be redefined using a value within the 0-to-10 scale. A score of 0 can be assigned to the minimum acceptable level of benefit, a score of 10 assigned to the highest level, and proportional scores between 0 and 10 assigned to the other benefit levels (Rothstein and Burna, 1997).

Table 4.6 presents an example set of scores for the set of example trading options defined in Table 4.3 and the decision criteria defined and weighted in Tables 4.4 and 4.5, using the common performance scale of 1 to 10 outlined above. The scores presented in Table 4.6 reflect the example WWTP's and its stakeholders sense of how the five alternative ways of meeting a 35% reduction in current loadings compare to each other along the seven criteria the WWTP established. In practice, a WWTP would develop a narrative definition of each possible score for each criterion and the participants in this exercise would use those definitions to assign scores. The definitions are not presented in this example.

Using a scale of 1 to 10, with 10 being the best, and 1 the worst, the example WWTP ranked options involving its facility higher on certainty and liability or risk factors, but lower on stewardship and partnership opportunities. Trading options involving other WWTPs, in this example situation, provided few benefits over nontrading options. This might be because there are few point sources with which to trade or because they face similar treatment costs. As has been seen in the Connecticut Long Island Sound, Neuse River, and Tar-Pamlico trading programs, cases exist where point–point trading can be very cost effective (Burkhart, 2003; Moore et al., 2000).

Options involving nonpoint sources offer advantages in minimizing overcompliance because of more flexible scaling of projects and in broader watershed benefits and partnerships.

A WWTP could stop here and make a visual comparison of the scores, consider background information and data that served as the basis for these scores, and move forward on the basis of those results. However, using a numerical scoring system in the manner described and illustrated above allows the WWTP to take the comparative analysis at least one step further and calculate the resulting total scores. This can be easily accomplished in simple tables or spreadsheets. Depending on the complexity of its options and evaluation criteria and the number of times the WWTP plans to conduct the evaluation, the WWTP may want to set up a spreadsheet model to facilitate conducting and reviewing the evaluation (Beaudet et al., 2001). The graphic representation of the scoring and ranking results for the example in Table 4.6 is presented in Figure 4.1.

By going through an exercise where the scores are multiplied by their weights and tabulated to maintain the results on a 1-to-10 scale, it is possible to calculate and compare the total scores of the evaluated options. For the example WWTP, some facility optimization appears to be the best near-term option. Facility investment plus a significant reliance on nonpoint-source credits to meet the 35% reduction target is a close second, but it may take longer to implement that option than some in-plant adjustments that could be done more quickly. Based on these results, however, using nonpoint-source credits could be a good option over the medium-term also, and so it will be worthwhile for the WWTP to explore these options more fully. The nontrading case involving incremental upgrades scored in the middle of the group, but, over time, this may prove to be the most expensive option overall—even more expensive than going to the limit of technology all at once. As noted in Table 4.6, in this case, point-source credit opportunities did not appear to be cost-effective options and also did not score well on some of the WWTP's other priority criteria. This also could change over time, as other point sources begin exploring their own trading options and optimization capabilities; thus, this option may be worth revisiting in the future.

Following the approach outlined in this section, the top-ranked results should reflect the WWTP's best values. For example, a WWTP may decide to implement a higher-cost option if the decision results in greater achievement of other goals and objectives of the WWTP. If the WWTP is lucky, it could end up with the best-value options being among the low-cost alternatives. Following the approach outlined here would help the WWTP compare results and select among competing scenarios. The MUA-based approach provides a systematic basis for comparing options that provide multiple benefits and helps to focus the analysis on those factors that should

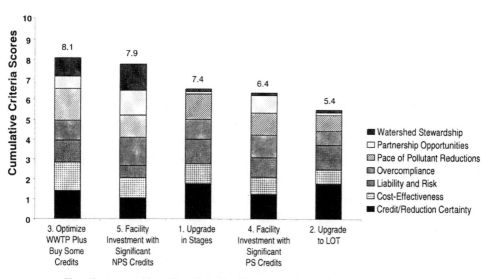

FIGURE 4.1 Total Scores and Overall Ranking. These example results were tabulated using the spreadsheet tool provided in AwwaRF's Capital Planning Strategy Manual (Beaudet et al., 2001). The format of the graphic output was modified outside of the tool for black and white presentation.

have the most influence on the WWTP's decision to trade or not to trade (Beaudet et al., 2001; Rogers et al., 1997).

Develop Trading Proposals

As stated at the outset of this chapter, the framework presented here to help the WWTP evaluate the benefits of trading versus not trading is necessarily general and best applied in an iterative decisionmaking process. It is general in the sense that the specific permutations and combinations of technical circumstances and local preferences are far too many to address in more detail. Additionally, it should be evident from the presentation of the framework and the example developed that a WWTP could benefit from using this framework at many stages in the trading process, including initial screening of potential opportunities, further development of trading (and nontrading) alternatives, and detailed development of trading proposals and the final decisionmaking process.

The framework takes a broad approach in evaluating the costs and benefits that may be associated with any trading option; this is in response to many recent experiences where a WWTP's decision about trading has not been made on the basis of cost-savings or cost-effectiveness alone. Instead, a host of other considerations also carry weight. Sometimes, these considerations can be monetized and integrated to the cost-benefit calculations and comparisons. However, more often than not, other priority considerations cannot be, or are not appropriately, monetized. For these, the WWTP and/or its stakeholders frequently demand the ability to identify these attributes of trading options in a transparent and direct fashion, so that everyone can evaluate trade-offs that may exist among alternative management strategies and trading options. This framework offers that capability.

At the end of each application of this evaluation framework, the WWTP will have identified the most promising options from among those evaluated. The next step may be to proceed with one or more of the best alternatives. The next step could also be to go back a step (sometimes, this type of analysis reveals data gaps, new questions, or even a surprising result) to reconsider certain aspects of the options or criteria and make some adjustments before reevaluating them.

Additional and more detailed discussion about many of the technical, regulatory, and implementation-oriented considerations related to evaluating and developing trading opportunities are found in other chapters, where there is more information about considerations that directly influence the specific costs and benefits that a WWTP will need to characterize for inclusion in the evaluation framework. These include source baselines and what might be tradable (Chapter 3); how to find potential trading partners (Chapter 5); the regulatory, institutional, and administrative arrangements (Chapters 5 and 7); and the public perceptions and stakeholder preferences that can play a significant role in determining overall feasibility of trading (Chapter 8).

References

Bacon, E. F.; Pearson, C. N., Jr. (2002) *Nitrogen Credit Trading in Maryland*; Water Environment Research Foundation: Alexandria, Virginia.

Bacon, E.; Rogers, J.; Ajello, T.; McElwaine, A. (2003) Multicredit Trading in a Pennsylvania Watershed. *Proceedings of the 76th Annual Water Environment Federation Technical Exposition and Conference*, Los Angeles, California, October 11–15; Water Environment Federation: Alexandria, Virginia.

Beaudet, B.; Bellamy, B.; Matichich, M.; Rogers, J.; Toukeda, A.; Wammock, G. (2001) *Capital Planning Strategy Manual*; 1P-5C-90838-5/01-ADS; American Water Works Association Research Foundation: Denver, Colorado.

Burkhart, M. (2003) Lower Neuse River Basin Association—Watershed Permitting to Increase Efficiency and Facilitate Trading. Presented at the National Forum on Water Quality Trading, Chicago, Illinois, July; U.S. Environmental Protection Agency: Washington, D.C.

Endangered Species Act (1973) U.S. Code, Chapter 35, Title 16.

Kieser, M. S. (2000) *Phosphorus Credit Trading in the Kalamazoo River Basin: Forging Nontraditional Partnerships*; Water Environment Research Foundation: Alexandria, Virginia.

Moore, R. E.; Overton, M.; Norwood, R. J.; DeRose, D. (2000) *Nitrogen Credit Trading in the Long Island Sound Watershed*; Water Environment Research Foundation: Alexandria, Virginia.

Paulson, C.; Vlier, J.; Fowler, A.; Sandquist, R.; Bacon, E. (2000) *Phosphorus Credit Trading in the Cherry Creek Basin: An Innovative Approach to Achieving Water Quality Benefits*; Water Environment Research Foundation: Alexandria, Virginia.

Rogers, J.; Burna, D.; Velicer, M.; Rothstein, E. (1997) *Decision Solutions: Decision Facilitation Guidebook*; CH2M Hill, Inc.: Denver, Colorado.

Rothstein, E. P.; Burna, D. (1997) Prioritization of Water Utility Capital Spending: An Analysis Framework. *Proceedings of 1997 AWWA Annual Conference*, Atlanta, Georgia; American Water Works Association: Denver, Colorado.

Rothstein, R.; Kiyosaki, D. (2003) Development of a Strategic Plan: Portfolio Management for Public Utilities. *J. Air Waste Manage. Assoc.*, **Jan.**

Safe Water Drinking Act (1974) Public Law 104-182; Code of Federal Regulations, Title 42.

CHAPTER FIVE

The Trade

Introduction

A statement commonly heard about water-quality trading is, "It sounds simple, but the devil is in the details". There is much truth to this observation, and many of the devilish details do not fully emerge until a regulatory agency or wastewater treatment plant (WWTP) begins to grapple with exactly how a trading program would work. This chapter addresses these devilish details: identifying trading needs, finding trading partners, structuring trades, instruments for trading, and permitting issues. Also covered are related issues, such as special considerations for point source–nonpoint-source trading and compliance and enforcement considerations.

The regulatory agencies have their own set of issues that are related to those covered in this chapter. They must establish the rules for trading, provide monitoring and oversight to ensure that trades are carried out within the rules and have the desired water-quality effects, track trading activity, take enforcement actions when appropriate, and assess the efficacy of trading programs. While these topics are of more interest to the regulatory agencies than to the WWTP, it is important for the WWTP to understand them and how they might affect what is required of the WWTP if it engages in trading. Chapter 7 is devoted to these issues; it discusses oversight considerations from the perspective of what regulatory agencies and society in general expect of trading programs.

General Trading Considerations

IDENTIFYING AND QUANTIFYING TRADING NEEDS

A critical first step for a WWTP contemplating trading, whether as a supplier of credits or a user, is to assess its current and future situations. A planning study of some type should be performed to understand all of the implications of trading. At a minimum, a WWTP should answer the following questions (Chapter 4 addresses the economics of trading; hence this discussion of the planning issues assumes that the economics are favorable for trading. In reality, the economic and technical-feasibility analyses would overlap and should be done concurrently):

■ What is the mass-load allocation for the pollutant of interest for the WWTP? This can be either a load limit in its National Pollutant Discharge Elimination System (NPDES) permit or a wasteload allocation under a total maximum daily load (TMDL).

■ What performance, with regard to the pollutant of interest, is the WWTP capable of at current wastewater flow? How reliable is this performance?

■ How are wastewater flows and loads projected to increase in both the short- and long-terms?

■ What would be the WWTP's best possible removal performance if actual flows were at design capacity?

■ When would plant expansion or upgrading be necessary?

■ In the long run, how would the WWTP expect to accommodate all of the projected growth in its service area?

Wastewater treatment plants contemplating being a supplier of credits should also answer the following questions:

■ Would the WWTP be able to reduce its discharges sufficiently below the level required by its permit or wasteload allocation to supply credits to other dischargers?

■ What is the estimated number of credits the WWTP could make available each year during the period in which it is able and willing to provide credits to others?

■ How reliably could the WWTP produce the credits over both the short- and long-terms?

■ How long would the WWTP be able or willing to generate credits to supply to others before needing its full treatment capability to meet its own needs?

Wastewater treatment plants contemplating using credits generated by others should also answer the following questions:

- Would the WWTP have to rely on credits supplied by others, or would it have the capability to meet its own needs if it wanted to?

- What is the estimated number of credits the WWTP would need from other sources each year during the period in which it wants to be a user?

- How long would the WWTP want to obtain credits from other dischargers before upgrading its facilities to meet its needs itself?

These questions should be answered, keeping in mind that the quantities of credits exchanged could be affected by trading ratios required by the trading program rules. Credits will not necessarily be exchanged on a one-to-one basis. Trading ratios are discussed in detail later in this chapter.

It is extremely important that the WWTP understand the long-term implications of trading. A trading arrangement that extended only through the next permit cycle could leave the WWTP in a precarious position if its situation changed in the following permit cycle or if its supplier or user of credits decided to end the arrangement. Careful consideration should be given to the duration of trading arrangements, and WWTPs must ensure that they can reliably meet their needs in the future, even if the current trading arrangement is no longer available. Additional questions that the WWTP should answer are as follows:

- How long would the trade arrangement last?

- What would happen when the trade expires?

- How would the WWTP meet its permit requirements?

  —Would there be a compliance schedule? How long?

  —Could the WWTP build the necessary facilities in time?

In one sense, these questions of duration imply that, in many circumstances, trades should not be considered permanent. A WWTP using trading to accommodate increases in its wastewater flow should be very cautious about allowing growth in its service area beyond its ultimate ability to comply with its permit without reliance on others. A WWTP could ignore this caution if the state were to provide ironclad assurances that credits would always be available in the necessary quantities from some source in the watershed. Such assurances are likely to be rare in trading programs, however.

Lest this caution seem too negative, it should be interpreted as merely illustrating that trading undoubtedly has limitations in the long run. Few, if any, trading programs could indefinitely compensate for sustained growth. (In fact, the whole issue of the effects of TMDLs and wasteload allocations on growth has largely gone unaddressed, so this is not merely a trading issue.) That does not mean, however, that trading cannot provide cost savings and other benefits in the short-term and probably for decades, in many cases.

The time frame in which credits must be generated and used should also be addressed in the planning analysis. The time frame should be specified in the trading program rules established by the state and should coincide with the time frame used in the analytic framework that established the load limits or allocations. For nutrients, TMDLS are frequently expressed in average annual loads. Hence, the typical time frame for nutrient trades is the calendar year (or seasonal, in the west), and credits must be used in the same year they are generated. In some cases, credit banking may be allowed, where credits produced in one year may be used in another, with a limit on how long a credit may be banked before it is used. Most pollutants do not lend themselves to banking in this manner, however, so such banking options will probably remain rare.

ELIGIBILITY TO TRADE

The U.S. Environmental Protection Agency (U.S. EPA) trading policy contains a single criterion for eligibility to participate in a trading program. It states that "sources and activities that are required to obtain a federal permit pursuant to Sections 402 or 404 of the [Clean Water Act] must do so to participate in a trade or trading program" (U.S. EPA, 2003a). Section 402 establishes the requirement for a discharger to obtain an NPDES permit, and Section 404 addresses dredge and fill permits issued by the U. S. Army Corps of Engineers. Because all WWTPs have been issued NPDES permits, this eligibility requirement is not an issue.

Additional eligibility requirements may be established by states in their trading programs. For instance, the *Chesapeake Bay Program Nutrient Trading Fundamental Principles and Guidelines* (U.S. EPA, 2001) lists, as one of the fundamental principles, that "traders must be in substantial compliance with all local, state, and Federal environmental laws, regulations, and programs". This provision was added to satisfy the environmental groups participating in the development of the principles and guidelines who were concerned that trading would be of great interest to so-called "bad-actors" who were merely seeking to evade their responsibilities. U.S. EPA's approach to substantial compliance is not so strict. The trading policy states that U.S. EPA "recommends that states and tribes consider the role of compliance history in determining source eligibility to participate in trading" (U.S. EPA, 2003a).

A WWTP investigating trading should not only ensure that it is eligible to trade, but that the potential trading partners, both point-source and nonpoint-source, are also eligible.

FINDING TRADING PARTNERS

After a WWTP identifies its trading needs, the next step is to find potential trading partners in the watershed. In many cases, this should be a simple and straightforward step; in others, it could become more complicated because of factors, such as the size of the watershed or the nature of the pollutant sources.

Types of Potential Trading Partners

There are two time-honored categories of pollutant sources: point and nonpoint. Both categories can be further delineated into subcategories. Point sources can be subdivided into the following categories:

■ WWTPs;

■ Industrial dischargers;

■ Concentrated animal feeding operations (CAFOs); and

■ Urban stormwater.

Urban stormwater is included in the point-source category because it is defined by the Clean Water Act (CWA) as a point-source discharge requiring an NPDES permit.

Nonpoint-source pollution is a broad category that essentially encompasses everything that is not considered point-source pollution. Types of nonpoint-source pollution include the following:

■ Agriculture-related water pollution. This category includes all agricultural-related activities, except for those requiring NPDES permits (e.g., CAFOs);

■ Nonagricultural-related water pollution. This category encompasses all nonagricultural pollutant loadings from rural and undeveloped lands. Loadings can be naturally occurring, such as nutrient loads in runoff from forested lands, or generated by human activity, such as silvaculture; and

■ Air deposition.

Where to Look for Potential Point-Source Trading Partners

If there is a TMDL for the watershed for the pollutant of interest, it is the obvious starting point in the search for potential trading partners. In most states, draft and final TMDLs are available for downloading on the state regulatory agency Web site. The TMDL should contain a list of all permitted point-source dischargers in the watershed, along with their discharged loads. Permit holders for NPDES stormwater permits and the stormwater-related loads should be identified.

If no TMDL exists, the state regulatory agency can provide lists of NPDES permit holders in the watershed, although it may not be able to readily provide information on the nature of the discharges. The WWTP may have to visit the agency and spend time looking through copies of NPDES permits to identify good potential point-source trading partners. If a trading program is already in existence, then this type of work has probably been done in the creation of the program.

Some states have active associations of municipal wastewater agencies (e.g., the Virginia Association of Municipal Wastewater Agencies). Such associations are valuable sources of information on all issues affecting WWTPs and would be a convenient way to meet and talk with potential trading partners. (As a side note, if a municipal wastewater organization exists in its state, it would be of great benefit for a WWTP to join it.)

Where to Look for Potential Nonpoint-Source Trading Partners

Because of the widespread and diffuse nature of nonpoint-source pollution, potential nonpoint-source trading partners should be more numerous and widespread. However, they will also probably be more difficult to find than point-source trading partners, and the WWTP will probably need to find more of them. Additionally, in some cases, it may be more effective to communicate with them through middlemen.

Agriculture will probably provide the most likely candidates and, fortunately, there are several state and local agencies and private organizations that provide support to farmers. These agencies and organizations would be good starting points, and it would probably be best to start at the most local level possible.

■ Soil conservation districts are a good starting point. Created by congress in 1938, there are approximately 3000 of them nationwide. They go by different names in some regions of the country (e.g., conservation districts and soil and water conservation districts) and distinctive names in some states (e.g., resource conservation districts in California and land conservation departments in Wisconsin). The boards of directors and supervisors of the districts are often farmers, and the districts work closely with the farmers in their district to develop and implement soil and other resource conservation plans. The soil conservation district office will know what farmers are doing, can probably provide insight to the best ways to approach them, and, perhaps most importantly, assist in various ways in facilitating trades.

In fact, the National Association of Soil Conservation Districts (NASCD) has concluded that "establishing a water-quality trading program to operate as a bank for nonpoint credits within a state agricultural cost-share program would…present the best format for channeling trading revenue to conservation on the land" (NASCD, 2003).

Extension services also have close contact with local farmers. Extension services are generally provided by the state agriculture departments, through the agricultural colleges of state land grant universities. They typically operate on the county level and assist farmers with both voluntary and regulatory programs. In Maryland, for example, the University of Maryland Cooperative Extension Service maintains offices in all counties in the state and staffs these offices with, among others, a nutrient management advisor. The extension services could provide the same sort of information and advice to the WWTP that the soil conservation district offices could.

■ State agricultural departments are another potential source of information and data. However, because they deal more with aggregated data and policy, it may be more difficult to obtain the type of information needed, and, in fact, the agricultural departments are likely to refer the WWTP to the extension services and soil conservation districts.

■ Irrigation districts exist in many regions of the country, notably in the west, and they have characteristics of both agricultural organizations and water utilities. Their purpose is to provide water for irrigation and, in many cases, for rural water supply. Most of them have also incorporated environmental considerations to their mission statements and attempt to operate in an environmentally responsible manner. Irrigation districts can also be regulated under the CWA, in some ways, and, hence, are likely to be active stakeholders in watersheds with water-quality impairments.

Nongovernmental organizations that could be extremely useful are the state and county chapters of the American Farm Bureau. The Farm Bureau is a private, nonprofit organization whose membership is largely comprised of farmers. The state and county chapters undertake a wide variety of activities to benefit farm families, including grass-roots policy development, legislative lobbying, publication of newsletters, and various family-oriented programs. Enlisting the Farm Bureau in bringing information about trading to farmers and attending Farm Bureau meetings and events would be an excellent way to make contact with potential trading partners in the agricultural community.

Beyond the agriculture sector, there could be a wide range of activities and organizations dealing with water-quality issues. Non-point-source loads from local, county, state, or federal lands could be suitable for trading arrangements. For example, as part of the Rahr Malting Company phosphorus offset trades described in Chapter 2, Rahr purchased a perpetual conservation easement for certain public lands from the city of New Ulm, Minnesota, and established vegetative cover and conservation practices on those lands to protect soil and water quality and produce the desired biochemical oxygen demand offset (Fang and Easter, 2003).

Third-Party Sources

There is great potential for third parties to play active roles in trading through various means. Some examples are as follows:

- Brokers or companies functioning as middlemen could contract with farmers or other sources for pollutant-reduction measures, and then package these reductions as credits for sale. Even government agencies, such as the Natural Resources Conservation Service of the U. S. Department of Agriculture or soil conservation districts, could conceivably perform this function.

- A WWTP could issue a request for proposals to purchase a certain number of credits for a certain number of years and award the contract to the lowest responsible bidder.

- A private company could function as a trade facilitator, providing the means to bring buyers and sellers together, regardless of which sector they are from, and then provide various support services for the trades.

- Commodity exchanges or state clearinghouses.

The internet could also become a valuable tool for finding trading partners and developing trades. The World Resources Institute (Washington, D.C.), an early proponent of water-quality trading (Faeth, 2000), is developing an Internet-based trading tool called NutrientNet (http://www.nutrientnet.org). According to its Web site, NutrientNet is designed to serve the following functions:

- Provide potential market participants and other stakeholders with background information on nutrient trading;

- Provide farmers, municipal treatment works, and industrial plants with tools for estimating releases of nutrients to surface waters from their operations, exploring reduction options, and estimating the costs of achieving reductions;

- Help market participants identify potential trading partners;

- Track the volume and type of trades within a watershed;

- Share lessons learned about trading across the watersheds where it is being tried or considered; and

- Provide information on water quality problems and trading as a possible means to address them.

The Web site is divided into the following three sections:

(1) Background information on nutrient trading.

(2) Worksheets for farmers and point sources to use to estimate their current nutrient loads, costs of achieving load reductions by different management practices, and the number of credits that could possibly be generated by the various management options.

(3)   A market section where offers to buy or sell credits can be posted. Summaries of market activities are also provided for each watershed.

An Internet-based system, such as NutrientNet, could also serve as a trading registry and help provide public transparency in trading programs. As of this writing, NutrientNet is in the prototype stage and the databases required for its operation have been fully developed only for the Kalamazoo River Basin in Michigan. The World Resources Institute is in the process of developing them for the Potomac River Basin in cooperation with the Chesapeake Bay Program.

While systems such as NutrientNet may be valuable in the future, none are currently available for use.

SPECIAL ISSUES WITH POINT–NONPOINT-SOURCE TRADING

Many observers have noted that, to date, there have been few true point-source–nonpoint-source trades. This observation has led some to conclude that marketlike point-source–nonpoint-source trading simply is not viable in the real world. Part of this negativism seems to stem from an unrealistic expectation that water-quality trading was to be the main way to finally achieve water-quality goals. In reality, this can never be the case, for many reasons; water-quality trading is just one tool to use when appropriate, and no one should be misled into thinking that it is a water-quality panacea.

Certainly, there are a number of policy issues and technical problems to be worked through in setting up point–nonpoint-source trading programs, and WWTPs are right to be wary. They know little about agriculture or the science and techniques of controlling agriculture-related water pollution (King and Kuch, 2003). (Note that if politics is added to the last statement, regulatory agencies do not seem to be in a very strong position either.)

Despite the technical and regulatory problems with rural nonpoint-source pollution, most observers, including the authors of this book, are more optimistic that the full potential of this type of trading will ultimately be realized. As more and more TMDLs place tight allocations on point sources for nutrients and other pollutants, threatening the ability of municipalities to grow, the demand for non-point-trading programs will increase. In discussing the lessons learned from the Dillion Reservoir phosphorus trading program, Richard Woodward wrote:

> Water pollution markets may be far from ideal, generating few and infrequent trades, but the option of trading has the potential to offer substantial benefits: improving the environment, allowing for economic growth, and benefiting many in the community (Woodward, 2003; p 11).

This section describes the technical issues that make nonpoint-source trading more challenging and addresses the steps the WWTP should take to protect itself in this type of trading. Much of the following discussion is based on the analysis and insight contained in a paper by Dennis King and Peter Kuch entitled Will Nutrient Credit Trading Ever Work? An Assessment of Supply and Demand Problems and Institutional Obstacles (King and Kuch, 2003). This paper is a highly recommended reference for its in-depth analysis of the apparent obstacles to point–nonpoint-source trading.

Quantifying Nonpoint-Source-Pollutant Loads and Reductions

In general, trading markets require that the units of trade be standardized (pounds of total nitrogen per year, for example). The unit of trade should have the same water-quality effect, no matter who and where the buyers and seller are. The public (and the regulatory agencies) have the right to expect that when precisely quantified and verifiable point-source-load reductions are replaced by nonpoint-source-load reductions, the net effect on the environment will be the same or better. The difficulty is that point-source loads are easily measured and verified and nonpoint-source loads are not; further, even estimating or modeling nonpoint-source loads is fraught with uncertainty.

For example, the phosphorus load originating from a crop field in any given year would depend on soil type, slope, crop type, fertilizer application rate, cultivation technique, use of cover crops, site hydrology, weather, and lag time before the phosphorus is delivered to the water body. Many of these factors would also vary from year to year, some randomly and some under the control of the farmer. Measuring or predicting the load would be fraught with difficulty, and the results would have a large degree of uncertainty (Lanyon, 1998). (Note that nitrogen would behave differently than phosphorus. Phosphorus loads are mainly associated with sediment and are delivered through surface runoff; nitrogen is soluble and its fate and transport once deposited on the land surface varies with the nitrogen species.)

It is important to note that this issue is a water-quality management problem, in general, and is not specific to trading. The issue comes to the forefront in trading programs because of the fundamental issue of the wisdom of trading measured, certain point loads for difficult-to-measure, uncertain nonpoint-source loads, and the challenges of structuring such trades.

Just as estimating pollutant loads is difficult, estimating the performance of agricultural BMPs is difficult for the same reasons.

King and Kuch (2003) presented a comprehensive list of farm management practices that "may result in tradable nonpoint-source nutrient credits." The list is as follows:

- Animal waste management (i.e. ponds, lagoons, tanks);
- Conservation tillage (e.g., no-till, low-till);
- Cover crops;
- Nutrient management;
- Retirement of highly erodable land;
- Runoff control;
- Erosion control;
- Stream protection with fencing;
- Stream protection without fencing;
- Forest conservation;
- Forest harvesting practices;

■ Forested buffers;

■ Grassed buffers;

■ Nonstructural shore erosion control;

■ Tree planting;

■ Enhanced stormwater management;

■ Erosion and sediment control (regulatory);

■ Stormwater management conversion; and

■ Stormwater management retrofits.

Best management practice performance is also very site-specific (Dillaha, 1998a, 1998b), and the load reductions achieved depend on many factors, including weather. Hence, BMP performance and reliability is somewhat difficult to generalize across the United States. Best management practices operate differently and are more predictable in the arid, irrigated west than they are in areas of the country with much higher rainfall (Schary, 2000). Fortunately, agricultural nonpoint-source loads and BMP performance are areas of active research and are a high priority for many agencies, including the United States Department of Agriculture (Gray, 2003) and others.

This uncertainty in nonpoint-source loads creates difficulty for both the trading partners and regulators that oversee it (King and Kuch, 2003). The problems can be summarized as follows:

■ It is difficult to define the necessary standard unit of trade;

■ Regulators must use complex scoring criteria to evaluate trades;

■ Credit buyers face risk that the seller will not perform as promised, or that insufficient reduction will be generated;

■ The uncertainty will result in increased transaction costs (e.g., cost of increased demands on the regulators or buyer cost associated with risk); and

■ Credit buyers face increased regulatory risk if they are held accountable by the regulatory agencies for the performance of the seller.

States will address the environmental risk associated with non-point-source trading and will incorporate measures to reduce it in the development of trading programs. Reducing the risk will invariably add costs to trading. The trick will be to balance risk reduction with cost so that point–nonpoint-source trading will be viable.

Trading Baselines for Farmers

The concept that there must be baseline requirements to be met and then exceeded by dischargers before credits can be generated to sell seems simple and logical. It is also a fundamental criterion by which trading programs are judged. (See the discussion of "additionality" in Chapter 7.) A WWTP with a wasteload allocation under a TMDL, for example, would have to decrease its discharged load below this allocation to generate credits to sell. For nonpoint sources, however, the TMDL load allocations are not assigned to individual sources like farms (there is little or no regulatory authority to do so) but are instead generalized, programmatic, and largely voluntary for individual farmers. That raises the question of which requirements a specific farmer must meet before being able to sell credits. This question is important to the WWTP seeking to buy credits from a nonpoint source because the answer would affect the availability, reliability, and cost of the credits.

U.S. EPA's Water Quality Trading Policy (2003a) defines "credits" as "reductions greater than those required by a regulatory requirement or established under a TMDL". This implies that, in cases where a TMDL containing a load allocation for nonpoint sources is in effect, nonpoint sources wishing to sell credits must first reduce their discharges below the load allocation. U.S. EPA has left it up to the states to decide how to define this and ensure that it occurs (Hall, 2004). In cases where there is no TMDL, U.S. EPA's position is that the baseline for nonpoint sources should be "the level of pollutant load associated with existing land uses and management practices that comply with applicable state, local or tribal regulations" (Hall, 2004).

Nationally, there is no consensus opinion on how to accomplish this, and the question has been answered differently in different trading programs. The Chesapeake Bay Program Nutrient Trading Fundamental Principles and Guidelines state

If state or Federal funds are used to cost share nutrient controls that generate credits, only that portion of those credits not paid for by the state or Federal cost share are available for trading (U.S. EPA, 2001; p 20).

This statement reflects the view that taxpayers have purchased pollutant-load reductions through agricultural cost-share programs and that these reductions cannot then be sold for profit by the recipient of the state and federal cost-share funds, a sale that would, in reality, negate the pollution reduction purchased by the taxpayer by allowing increased discharges by the credit purchaser. Not everyone agrees with this baseline restriction, however, especially in the agricultural community, and there is no national consensus on this issue.

The baseline requirements proposed in the Water Quality Trading regulations adopted by the Michigan Department of Environmental Quality in 2002 (Michigan Department of Environmental Quality, 2002) are that farmers wishing to sell nutrient credits must be in compliance with a set of generally accepted agricultural and management practices (GAAMPs) developed by the Michigan Commission of Agriculture (1997a, 1997b, 1997c). These GAAMPs deal with nutrients, manure, and pesticide management, respectively, and were developed to shield farmers who comply with them from nuisance lawsuits.

The issue of agricultural baselines is not of direct concern to WWTPs. It is a state's responsibility to define them in the development of the trading program. The WWTP should understand, however, how the baseline requirements will affect the availability and cost of credits. In general, the existence of baselines will both decrease their availability and increase their cost (King and Koch, 2003); hence, unnecessarily high baseline requirements could reduce the viability of trading.

Solving the Nonpoint-Source Uncertainty in Trades

U.S. EPA's Trading Policy states that "standardized protocols are necessary to quantify pollutant loads, load reductions, and credits" and "where trading involves nonpoint sources, states and tribes should adopt methods to account for the greater uncertainty in estimates of nonpoint source loads and reductions" (U.S. EPA, 2003a). The policy lists several possible approaches

- ■ Direct monitoring;
- ■ Trading ratios;
- ■ Use of demonstrated performance values or conservative assumptions in estimating effectiveness of nonpoint-source management practices;
- ■ Site- or trade-specific discount factors; and
- ■ Retiring a certain percentage of nonpoint-source reductions for each transaction.

Direct on-site monitoring of nonpoint-source loads and BMPs, as noted earlier, would be extremely difficult and expensive to do. Hence it has rarely, if ever, been done in trading programs. The most common method in practice, thus far, is the use of a trading ratio to account for uncertainty. Uncertainty ratios are only one of several types of trading ratios that could be applied to a single trade. Note that the last two methods on U.S. EPA's list are, in effect, applications of trading ratios. A complete discussion of trading ratios is presented in the next section of this chapter.

Another approach would be for the state to define standardized nonpoint-source loads and BMP reductions. In assessing how to best estimate nonpoint-source nutrient loads for the Chesapeake Bay Program, Lanyon concluded that "pursuing a rule-based approach that does not require specific monitoring results may have some utility in real-world situations" (Lanyon, 1998). The approach uses research results and literature values to estimate pollutant loads and the reductions produced by various BMPs, and the selected values are defined in state policy or regulation. In the Idaho phosphorus trading program, for example, the state intends to adopt a list of approved BMPs that was developed with stakeholder involvement (Ross and Associates, 2000). The performance of these BMPs has been established through sufficient data collection on their performance (Carter 2002; Idaho Soil Conservation Commission, 2002), and only flow data are needed to calculate the quantity of credits produced. An uncertainty factor is then added to account for random variability. It would also be possible to use "average" credit values to minimize the need to account for variability. (The full report of the Idaho Soil Conservation Commission is included in Appendix B.) Only BMPs on the approved list can be used to generate credits for trading.

As more research is done on BMP performance, uncertainty factors should decrease.

Avoiding Nonpoint-Source Uncertainty in Trades

One way for WWTPs to avoid the issues associated with nonpoint uncertainty is to purchase credits from the state by paying money to a state cost-share program of some sort, if the state has made this option available in the trading program. While not on the Water Quality Trading Policy list (U.S. EPA, 2003a), it is consistent with the requirement for standardized protocols and is already being used in trading programs. In North Carolina, the Tar-Pamlico Trading Association can purchase credits from the agricultural cost-share program, if it is unable to meet its allocation by other means. Likewise, the Neuse River Compliance Association can purchase credits from the North Carolina wetlands restoration fund.

Purchasing credits from the state has the advantage, from the WWTP point of view, of transferring responsibility for the uncertainty problem to the state. The WWTP is also freed from enforcement concerns or any liability for seller nonperformance. A possible disadvantage is that such credits may be relatively expensive to purchase compared with purchase from farmers or middlemen.

TRADING RATIOS

A trading ratio is a requirement that a buyer of water-quality credits actually purchase more credits than needed to meet its own discharge requirements. There are different types of trading ratios that are imposed for different reasons, and a single trade could have multiple trading ratio requirements placed on it, depending on the circumstances.

There are at least five types of trading ratios: uncertainty, delivery, water-quality, retirement, and cross-pollutant. Each is described in detail below.

Uncertainty Ratio

An uncertainty ratio is applied when actual pollutant loads or reductions cannot be accurately measured. In trades between WWTPs, this is not an issue—effluent volumes and pollutant concentrations are precisely measured and reported to the regulatory agency on a regular basis. A trade between WWTPs should have a trading ratio of 1:1, and the WWTP should resist any higher ratio unless a valid reason for it can be provided.

Point-source–nonpoint-source trades, however, will invariably incorporate trading ratios. The uncertainty inherent in nonpoint-source pollutant loads and the measures implemented to reduce them are described in detail in the previous section. The application of the uncertainty ratio provides a margin of safety to ensure that the actual loads resulting from a trade do not violate the water-quality requirements, despite the inability to precisely measure them.

There are two types of uncertainty inherent in nonpoint-source loads: a lack of knowledge of what the pollutant loads actually are under various conditions (mainly because of the difficulty of measuring or modeling them), and how they vary randomly because of weather and other factors. Typically, both types of uncertainty are covered by a single ratio. Hopefully, in the future, greater scientific understanding of the dynamics of nonpoint-source pollution will allow the use of lower uncertainty ratios.

Uncertainty ratios of 2:1 or 3:1 are typically required, so a point source needing 113 398 kg (250 000 lb) of phosphorus credits in a given year to meet its permit requirements may be required to purchase 226 796 kg (500 000 lb) or 340 194 kg (750 000 lb) of credits. Or alternately, the nonpoint source may be required to produce 226 796 kg (500 000 lb) or 340 194 kg (750 000 lb) of estimated reductions to sell the 113 398 kg (250 000 lb).

The Cherry Creek phosphorus trading program (Paulson et al., 2000), the Kalamazoo, Michigan, phosphorus trading program (Kieser., 2000), and the Lower Boise phosphorus trading program all use this type of ratio (Ross and Associates, 2000).

Cherry Creek Basin Phosphorus Credit Trading Program Use of Trading Ratios

Among its many activities, the Cherry Creek Basin Water Quality Authority constructed four pollution reduction facilities (PRF) directly adjacent to Cherry Creek Reservoir in Colorado. Three of the PRFs are designed to slow storm flows and baseflows in the streams feeding the reservoir. Two of these PRFs feature an initial retention pond and then wetland ponds in series; one is an extended infiltration facility. The fourth PRF is a restoration project on the reservoir shoreline.

These four PRFs generate reservoir phosphorus-load reductions that are available as credits that can be purchased by member jurisdictions. The authority devoted considerable effort to developing detailed and specific trading ratios to use throughout the watershed (not only for these four PRFs). Each trading ratio is composed of three factors.

(1) An institutional uncertainty factor based on the history and stability of the entity constructing and owning the BMP (in the case of the Cherry Creek Basin Water Quality Authority, 1.0);

(2) A variability factor derived by dividing the average annual phosphorus-load reduction achieved by a PRF by the lower 95th percentile value of the reduction; and

(3) A "best professional judgment" factor, made up of the following five elements:

■ Data limitations;

■ Age of the wetland ponds;

■ Location in the watershed;

■ Timing of load reductions; and

> ■ Phosphorus chemistry (i.e., soluble versus sediment-associated).
>
> The net trading ratios developed for the four facilities are 1.4, 3.0, and 1.7 for the three stormwater detention PRFs and 1.8 for the streambank restoration project. The precision of these ratios is a result of the extensive analytical effort made by the Authority. This effort also had the benefit of avoiding the imposition of unnecessarily conservative ratios (Paulson et al., 2000).

Delivery Ratio

When a pollutant is discharged to a water body, its effect on a given segment of the water body depends on many factors, such as physical, chemical, or biological activities that affect it; how it is transported; the existence of pollutant sinks; etc. For example, the percentages of phosphorus and nitrogen in a WWTP's discharge to a river that reaches a spot 160 km (100 mi) downstream will be different because of the difference in behavior of the two pollutants. Nitrates in the discharge can be converted to nitrogen gas by biological activity in the stream and lost to the atmosphere. The phosphorus in the discharge may be adsorbed to sediment particles, lost to bottom sediments, and transported at far different rates in the river than the soluble nitrogen.

When the buyer and seller of credits are located at different points in the watershed, a delivery ratio is used to account for the differences in delivery of the pollutant to the point in the water body requiring protection. This helps to ensure that the trade produces equivalent water-quality effects. The ratios (or attenuation factors, as they are sometimes called) are directly affected by the distances involved.

> *Connecticut Long Island Sound Nitrogen Trading Program*
> *Use of Delivery Ratios*
>
> The low dissolved oxygen problems in Long Island Sound occur in its western end. Connecticut dischargers are spread out among six zones, with zone 1 being at the eastern end of the state, and zone 6

TABLE 5.1 Delivery factors for the Connecticut zones.

ZONE	NAME	FACTOR
1	Thames River Basin	0.18
2	Connecticut River Basin	0.22
3	Quinnipiac River Basin	0.60
4	Housatonic River Basin	0.67
5	Saugatuck River Basin	0.85
6	Norwalk River Basin	1.00

at the western end, closest to the anoxia problem. Water-quality models of the sound and its watershed were used to establish "delivery" factors that account for the relative oxygen effect from a pound of nitrogen. A pound of nitrogen discharged closest to the area of impairment (zone 6) has a delivery factor of 1.0 (no assigned attenuation). The delivery factors for the Connecticut zones are shown in Table 5.1.

To account for the other component of attenuation (that which occurs in the rivers), Connecticut developed "tier delivery factors". The number of tiers ranges from one tier for small, coastal watersheds, where no added attenuation is assigned (tier factor = 1.00) to four tiers for the largest watershed. Tier factors decrease with distance from the sound, accounting for less nitrogen delivery efficiency (Table 5.2).

The state purchases and sells all credits. To do this, it applies trading ratios, calculated from the zone and tier delivery factors, to normalize all credits to a zone 6, tier 1 basis. This conversion creates trading credits based on a "normalized exchange rate".

To illustrate the application of the delivery factors to create trading ratios, if a WWTP in the Shetucket Tributary of the Thames River basin needs to buy credits to offset 454 kg (1000 lb) of nitrogen discharge, those 454 kg (1000 lb) are first converted to equivalent zone 1, tier 1 kilograms (pounds) (eq 1).

TABLE 5.2   Tier factors in Connecticut.

RIVER BASIN	TIER 1	TIER 2	TIER 3	TIER 4
Thames	1.00	0.91	0.75/0.83[a]	—
Connecticut	1.00	0.93	0.87	0.81
Quinnipiac	1.00	0.83	—	—
Housatonic	1.00	0.69/0.90[b]	0.52	—
Saugatuck	1.00	—	—	
Norwalk	1.00	—	—	—

[a]0.75 in the Quinebaug Tributary and 0.83 in the Shetucket Tributary.
[b]0.90 in the Naugatuck Tributary and 0.69 elsewhere.

Credits needed = 454 kg × (zone 1 factor) × (Shetucket tier
(1) factor)
= 454 kg × 0.18 × 0.83
= 149.4 kg (68 lb)

(Information courtesy of Paul Stacey, Connecticut Department of Environmental Protection [Stacey, 2004]).

Retirement Ratio

A retirement ratio is imposed to provide a water-quality benefit beyond that achieved by the trading partners merely meeting their assigned allocations. In a sense, it can be considered a "tax" on trades. The Michigan trading regulations require that 10% of all water-quality credits that are sold must be retired instead of used by the purchaser "to ensure net reductions and progress toward water quality goals" (Michigan Department of Environmental Quality, 2002).

Kalamazoo River Basin Phosphorus Trading Program
Use of Trading Ratios

The Kalamazoo trading program uses two ratios. The first is an uncertainty factor (called a discount factor by the program). Very specific discount factors were developed for each of the four nonpoint-source trading "sites" studied by the program. One of the sites, known as the Georgia Pacific site after the owner of the land, is an area of high streambank erosion. After careful study, the program determined that the most technically accurate estimate of annual phosphorus-load reductions that could be achieved is 36.3 kg (80 lb) per year, but that a more conservative estimate of 33.1 kg (73 lb) per year (or 91%) would be used to account for uncertainty. Therefore, each kilogram or pound of phosphorus credit sold is discounted by 9%.

A policy decision was also made by the program that a 50% retirement ratio would also be applied to all trades to ensure net improvement in water quality. This ratio is referred to by the program simply as the trading ratio.

The uncertainty ratio and retirement ratio are used together to quantify a trade. A sale of 454 kg (1000 lb) of phosphorus would result in the following (eq 2):

$$454 \text{ kg} \times 0.5 \times 0.91 = 206 \text{ kg (455 lb)} \tag{2}$$

Thus, 206 kg (455 lb) of phosphorus credits would be acquired by the buyer. Conversely, a buyer wishing to acquire 454 kg (1000 lb) of credits would have to purchase 997 kg (2198 lb), as follows (Kieser, 2000):

$$\frac{454 \text{ kg}}{0.5 \times 0.91} = 997 \text{ kg (2198 lb)} \tag{3}$$

Cross-Pollutant Ratio

In the rare cases where cross-pollutant trades occur, a ratio that equates the two pollutants must be developed. The most notable example is the Rahr Malting Company five-day carbonaceous biochemical oxygen demand ($CBOD_5$) offset described in Chapter 2. The Minnesota Pollution Control Agency (MPCA) evaluated the effects of $CBOD_5$ discharges and nonpoint-source phosphorus and nitrogen loads on dissolved oxygen levels in the impairment area downstream of the Rahr discharge, and the agency determined that the appropriate cross-pollutant trading ratio for $CBOD_5$:phosphorus:nitrogen is 1:8:4. Uncertainty and delivery ratios are also applied (MPCA, 1997).

For the most part, there are valid water-quality reasons for the imposition of trading ratios. They mitigate the uncertainty inherent in nonpoint-source pollution control, account for the variation in water-quality effect because of the location of the discharge, help achieve water-quality goals, and enable cross-pollutant trading. The downside is that trading ratios increase the cost of trading, possibly so much that trading becomes economically nonviable (King and Koch, 2003). The WWTP should try to ensure that the trading ratios applied by the state in its trading program are necessary and appropriate. Further, additional research on nonpoint-source loads and BMP performance should be supported. If uncertainty can be reduced, then the uncertainty ratios can be reduced. The Cherry Creek Basin Authority, for example, was able to use trading ratios as low as 1.3:1 because of its extensive study of the BMPs it installed (Paulson et al., 2000).

On the other hand, trading ratios can sometimes benefit WWTPs. Delivery ratios can introduce equity to trading programs by insuring that WWTPs buying credits are purchasing the amount of credits that exactly offset the water-quality effect of their own discharges. The Connecticut Long Island Sound nitrogen trading program is the most notable example of this.

A second way that trading ratios can benefit WWTPs is that the use of uncertainty ratios in point–nonpoint-source trades should shield the WWTP from issues arising from the actual performance of the BMPs in any given year. The WWTP, in essence, has purchased extra credits to mitigate the environmental risk of variable and uncertain BMP performance and, hence, cannot legitimately be held accountable for any variation in actual BMP performance. To do so would be making the WWTP pay twice for the same thing. Wastewater treatment plants should insist that their acceptance of the trading ratios free them of responsibility for actual BMP performance, assuming that the BMPs are properly constructed and maintained.

The WWTP should also be wary of the indiscriminant use of retirement ratios. This type of ratio is considered by some as a way to leverage additional pollutant reductions beyond that required of the trading partners to meet their allocation requirements. Inherent in this is a philosophy that pollutant reductions to levels below allocations are always beneficial (less is always better, all the way to zero). Scientifically, this is a debatable claim, although the CWA (1972) does envision an ultimate goal of "zero discharge" of pollutants. Another reason for this leveraging would be to help offset pollutant loads from sources that do not readily lend themselves to control. This, however, would constitute a shift of the pollution-control burden from the responsible source and onto another discharger, violating the principle that "the polluter pays" and raising serious equity issues. Unfortunately, it has not been uncommon to hear statements such as "water-quality trading is a way to bring point-source dollars to bear on nonpoint-source problems". The reaction of the WWTP to this should be "wait, those are my dollars but not my pollution. I'm willing to do my fair share, but...".

CREDIT FOR MULTIPLE BENEFITS

In many cases, a point–nonpoint-source trade would result in additional benefits beyond the reduction in the pollutant being traded. For example, a trade in which a WWTP purchased phosphorus credits from a nonpoint source could also result in BMPs and stream restoration and stabilization activities that would produce the following improvements:

- ■ A more natural flow pattern in the stream, resulting in improved aquatic habitat;
- ■ Reduction in sediment loads;
- ■ Reduction in streambank erosion;
- ■ Exclusion of livestock from the stream and streambank, resulting in lower bacteria loadings;
- ■ Creation of vegetated stream buffers; and
- ■ Shading of the stream by buffer-zone vegetation.

Both the terrestrial and aquatic habitats would clearly be improved by this trade in several different ways.

In reality, many streams are impaired by multiple causes (flow, physical habitat, sediment, etc.), and this should be kept in mind when designing and improving water-quality trades (as a watershed management approach would require). It would not be unreasonable for the WWTP to argue that ways should be sought to include these multiple benefits in trade calculations (i.e., Can these benefits be quantified somehow and credited in a trade? Can these benefits be used to offset what may be perceived as unfair or unnecessarily high trading ratios?). Rigidly applying multiple trading ratios to each parameter being traded may cause one to miss a larger picture about how trades could benefit streams.

At least one attempt has been made to gain credit for multiple benefits (see the Rahr Malting Company Phosphorus Trades and Multiple Benefits section below), and at least one researcher is attempting to develop the concept in a broader way (Kieser, 2003). Finding ways to document, quantify, and incorporate ancillary benefits to trades should be a high priority for U.S. EPA, state regulatory agencies, WWTPs, and anyone interested in a holistic watershed management approach to water-quality protection.

> ### Rahr Malting Company Phosphorus Trades and Multiple Benefits
>
> In its negotiations with the MPCA over the phosphorus trading arrangements described in Chapter 2, Rahr Malting proposed to MPCA that the wetland it was constructing to generate sediment and, hence, phosphorus-load reductions should also generate credits from the wetlands mitigation bank. The MPCA responded that it would be unacceptable to grant double credit for the same investment. Kerr et al. (2000; p 74) identified the following two separate questions in the Rahr request and the MPCA response: "Are there, in fact, two separate sets of benefits?" and "If a project design creates independent benefits, should credits be prohibited for more than one of those benefits?" Kerr et al. (2000) concluded that the first question could often be difficult to answer, but that "...dual-credits for dual benefits might provide a market incentive for more integrated pollution-reduction and/or mitigation measures", and that "there might be significant environmental and economic benefits to encouraging, in those cases where it is feasible, designs that could create multiple benefits" (Kerr et al., 2000; p 75).

WATER QUALITY REVISITED—LOCAL EFFECTS BECAUSE OF TRADES

A major concern frequently expressed about water-quality trading, especially by skeptics, is that trades could result in adverse local water-quality effects. The buyer of credits, in effect, imports additional pollutant loads into the local watershed. If the buyer discharges directly to the target water body that was the focus of the analytical framework that set the load allocations or permit limits, then there would be no issue with adverse local effects because of trades. However, if the target water body is remote from the buyer (for example, a reservoir 80 km [50 mi] downstream of the buyer's discharge), then the potential would exist for the trade to adversely affect the river in the intervening 80 km (50 mi). This potential effect on "local" water-quality must be evaluated in developing the trade.

In drafting the Water Quality Trading Policy, U.S. EPA was fully aware of this issue. The policy states

> [U.S.] EPA does not support any use of credits or trading activity that would cause an impairment of existing or designated uses, adversely affect water quality at an intake for drinking water supply or that would exceed a cap established under a TMDL (U.S. EPA, 2003a).

The Chesapeake Bay Program Nutrient Trading Fundamental Principles and Guidelines (U.S. EPA, 2001; p 15) also explicitly addressed the issue. The importance of the issue is reflected in the fact that it is fundamental principle 1

Trades must not produce water-quality effects locally, downstream, or baywide that

- ■ Violate water-quality standards or criteria,
- ■ Do not protect designated uses, or
- ■ Adversely impact living resources and habitat (U.S. EPA, 2001).

This approach goes somewhat beyond U.S. EPA's reliance on water-quality criteria and adds the more subjective and difficult-to-measure "impact on living resources and habitat".

At some point in developing the trade, the proposed post-trade loads must be assessed for unforeseen water-quality effects in all water bodies potentially affected by the trade. (As noted above, however, if the buyer discharges directly to the target water body, this should not be necessary.) This must be a routine part of the planning needed to develop trades.

Whether the analysis is performed by the state or the WWTP proposing the trade is a matter to be decided by these parties. It is generally the state's role, however, to assess the water-quality effects of discharges; hence, the state presumably possesses the staff and analytical tools needed to do so for the proposed trade. If insufficient monitoring data exists, the state may ask the WWTP to collect additional data.

The local water-quality effects issue has an interesting converse. If limits on a pollutant are set for local water-quality reasons, could this sometimes generate credits that could be used by more distant sources for a larger-scale water-quality initiative? An example of this would be a tidal tributary to the Chesapeake Bay for which a nitrogen TMDL has been adopted. If the TMDL allocation is lower than the larger-scale allocation assigned as a result of the Chesapeake Bay Program requirements, could the local dischargers sell the unusable portion of their bay allocation to other bay dischargers? This is an interesting question that would require a policy decision to resolve.

Protecting Local Water Quality from Adverse Effects Because of Trades—An Example

The Connecticut Long Island Sound general permit for nitrogen stipulates the total nitrogen-load allocations for all of the WWTPs in the state. Each WWTP also has its individual NPDES permit for all other requirements. These permits contain limits on ammonia-nitrogen, designed to protect the local receiving waters from ammonia toxicity. Regardless of the number of nitrogen credits a WWTP purchases so it can continue to discharge nitrogen, the ammonia limits must be met. Hence, local water quality is protected from any effects from the trade.

Trading Instruments

This section addresses the two major legal mechanisms that will be present in most trades: contracts between the users and providers of credits and the provisions for trading that states may include in the NPDES permits of one or both of the trading partners.

Both trading contracts and NPDES-permit conditions for trading are new developments in water-quality management and law. There is no body of established practice to rely on for precedents and very few real-world examples to study. Practices in these two areas will rapidly evolve as water-quality trading matures. For these reasons, this book cannot and should not attempt to be definitive in these areas. Regarding contracts, this chapter generally addresses the functions a contract should perform for the trading partners. On permits, it discusses the permit requirements and considerations contained in U.S. EPA's Water Quality Trading Policy (U.S. EPA, 2003a). The chapter then concludes with a discussion of ways in which WWTPs can minimize the potential risks involved with incorporating water-quality trades to discharge permits.

The following discussion neither attempts to prescribe what any particular contract or NPDES permit should contain, nor does it cover what would be considered standard contract language. Readers contemplating developing trading contracts or negotiating permit conditions should consult with their legal counsel and should not rely solely on this book.

CONTRACTS AND AGREEMENTS

Contracts will be critical for many types of trades. They are the instruments in which the users and providers of credits set forth the conditions of the trade and the obligations of the trading partners. They should also provide protection to both parties in the event the other defaults. A well-crafted contract provides both clarity and protection to both parties.

Contracts are very flexible and powerful instruments that can be tailored in form and substance to almost any circumstance. In recent years, contract-based approaches have been increasingly woven into state and federal water-quality programs. Examples of this trend include the Biological Nutrient Removal Agreements between Maryland WWTPs and the Maryland Department of the Environment in use during the 1990s and the Water Quality Improvement Grant Agreements used in Virginia since 1997.

The nature of the contracts that may be required will differ with the type of trading program (managed trading, trading association, or marketlike trading [small-scale offset programs would be very similar to marketlike trading regarding contract and permitting issues]).

Managed Trading

Managed trading programs do not require contracts between traders. Instead, the state establishes trading requirements in NPDES permits, either in individual WWTP permits or in a general, statewide permit.

Trading Associations

Thus far, only two trading associations have been created in the United States: the Tar-Pamlico and Neuse River associations in North Carolina. The Neuse River Compliance Association does not have a contract with the state; it operates under the authority and requirements of state regulations that were adopted for that purpose. That leaves the Tar-Pamlico Trading Association as the only trading association currently operating under an agreement negotiated with the state. The agreement is described here.

A Nutrient-Sensitive Waters Implementation Strategy: An Agreement between the Tar-Pamlico Trading Association and the North Carolina Department of Water Quality, Division of Environmental Management*, 1989

I.　**Background and Purpose**

II.　**Association Members.** The fourteen original members of the association are listed in this section. It also establishes conditions and timetables for adding new members.

III.　**Nutrient Reduction Targets.** This lengthy section contains a detailed history of the derivation of the water-quality goals and the nitrogen and phosphorus reduction goals assigned to the association, reading more like a technical report than an agreement. It also discusses nonpoint-source loads and obligates the Division of Environmental Management to work with appropriate groups and agencies to "establish a coordinated and focused plan to achieve the required nonpoint source reductions". Further, it states that

---

*Note: In 1989, the Department of Environmental Management was renamed the Department of Water Quality.

National Pollutant Discharge Elimination System permit limits of 6 mg/L total nitrogen and 1 mg/L total phosphorus will be established for non-association dischargers. It also addresses how expanded discharges and new facilities would be handled.

IV. **Trading Program.** The "trading program" established in this section provides the association with additional opportunities for meeting its allocation. Trading options listed include funding agricultural best management practices through state programs, funding state staff positions, supporting the development of nutrient management plans for non-agricultural sources, and funding the State Agricultural Cost-Share Program.

V. **Minimum Conditions to this Agreement.** The association agreed to perform effluent monitoring, submit an annual report, pursue federal funding for a nutrient fate and transport model, and provide funding for support of the estuary water-quality model.

VI. **Local Water-Quality Impacts.** This section states that additional nutrient reductions may be required in some locations to address local water-quality problems.

VII. **Decision-Making Authority.** The Division of Environmental Management's ultimate authority to make all decisions regarding nutrient allocations and tradeoffs is recognized in this section.

VIII. **Nonpoint Source Controls.** This section acknowledges that there are other nonpoint-source-control initiatives underway in the watershed.

IX. **Violation of Terms of this Agreement.** The strategy that would be implemented by the Division of Environmental Management in the event that the terms of the agreement were violated (presumably by the association) is delineated in this section. In essence, the requirements described in Section III for nonassociation dischargers would be imposed on association members.

It is interesting to note that Section III of the Tar-Pamlico Agreement obligates the state to impose permit limits on nonassociation dischargers and attempt to deal effectively with nonpoint sources. This accomplished two things; first, it took away the option for a discharger to elect to do nothing (neither join the association nor reduce nutrient discharges). Second, it forced recognition that water-quality goals could not be achieved in the basin without the implementation of substantial controls for nonpoint-source loadings.

The Tar-Pamlico agreement serves as a good starting point for developing a proposed association agreement. At a minimum, an agreement between an association and the state should include the following:

■ Background and purpose;

■ Association members (and provisions for adding or dropping members, if appropriate);

■ Water-quality goals;

■ Pollutant-load allocations;

■ Compliance schedules;

■ Obligations of the state;

■ Obligations of the association;

■ Provisions for resolving violations of the agreement; and

■ Duration.

Obligations of the state could include items such as providing grant or loan funds and ensuring that nonassociation dischargers and nonpoint sources do their fair share in the watershed (as in the Tar-Pamlico agreement). A provision that could be very valuable to an association would be one that provided an alternate (or emergency) source of credits, in the event the association found itself unable to meet its allocation in any given year. Another item of interest to WWTPs would be how the state would deal with proposed new dischargers in the watershed.

The obligations of the association could include items such as optimizing the performance of existing and new facilities and reporting. Other provisions addressing special circumstances in the state, watershed, or association should be included.

The duration of the trade is an important long-term planning issue for the trading partners. No matter what the duration—1 year or 30 years—it is critical for the WWTP to answer the question "What happens after the expiration of the trading contract?" before executing the trade. If a trade is a permanent part of a WWTP's plan, then the contract needs to address that in some manner. The planning questions posed at the beginning of this chapter provide the framework for a WWTP to assess its needs and options for the long run.

Marketlike Trading

Marketlike trading and small-scale offset programs will nearly always require contracts between users and providers of credits. The contract issues would be different for different types of trading partners (other WWTPs, individual nonpoint sources, and "middlemen") and are addressed separately.

Contracts with Other Wastewater Treatment Plants

When both the user and provider are WWTPs, the contract could probably be fairly straightforward and simple. As far as is known, there are no examples of this type of trading contract yet. At a minimum, the contract should include the following:

- Purpose of the contract;
- Quantities of credits exchanged;
- Prices of credits exchanged;
- Duration of the contract;
- Obligations of the seller (agreement to accept permit limits reflecting the trade);
- Obligations of the buyer (agreement to accept permit limits reflecting the trade); and
- Provisions for resolving violations of the agreement.

Other possible provisions could include options to escalate prices, if appropriate; extend the contract beyond the specified duration; or cancel the contract. Both parties would be adversely affected by a default of the provider of credits, so the WWTP using the credits should include a provision to address any potential defaults. This concern is covered in greater depth in the last section of this chapter.

Contracts with Individual Nonpoint Sources

Trades between a WWTP and individual nonpoint sources, such as farmers, would raise many more issues that would have to be addressed in the contracts. Further, contracting with multiple individual sources and monitoring them for compliance would produce additional resource demands and legal issues that should be addressed from the outset. At a minimum, the contract should include the following:

- Purpose of the contract.
- Quantities of credits exchanged.
- Prices of credits exchanged.
- Duration of the contract.
- Obligations of the seller, including an agreement to undertake specified actions to reduce pollutant loads; agreement to properly maintain BMPs or other specified facilities; agreement to allow regular inspections by buyer and/or third parties; and agreement to comply with all federal, state, and local requirements.
- Obligations of the buyer.
- Provisions for violation.

The duration of the contract is a more complicated issue than with trades between two WWTPs. If the trade requires that BMPs or other facilities be constructed, and the price of the credits reflects those costs, then the buyer should reasonably expect to acquire the credits for the full useful lives of the facilities. This could conceivably be many years for some types of facilities (detention ponds or constructed wetlands, for example). Failure of a BMP could raise the issue of whether the failure was a violation of the contract (i.e., because of factors beyond the reasonable control of the seller or failure to properly maintain the BMP).

The WWTP should also pay close attention to potential federal, state, and local requirements that may apply to the generation and sale of credits. For example, the state may require that all credits sold under a trading program must first be certified by the state. The state may also require that it have the right to inspect credit-generating facilities, or that credit generation must be independently verified by a state-approved third party. The WWTP should review all applicable federal, state, and local regulations or policies and address any such requirements. Consultation with state and federal regulators will be critical at the early planning stages until any local trading rules or policies are established and announced.

Both parties to the trade and the state may sometimes want to use third parties for certain activities. For example, a state farm bureau or extension service could play any number of roles in facilitating or verifying nonpoint-source-related trades. If so, the third parties and their roles in the trade should be stipulated in the contract.

If uncertainty ratios are not being used in the trade, then the contract would also have to include assignment of responsibilities for monitoring and reporting actual pollutant-load reductions. As discussed in the section on trading ratios, this approach would be very technically demanding and resource-intensive and is not recommended, except in special circumstances where monitoring of actual loads is practical and relatively inexpensive. While uncertainty ratios would increase the cost of the trade in one way, they also allow the avoidance of potentially high costs for monitoring and measuring pollutant loads on a continuous basis.

This type of trading "hassle-factor" would vary, depending on the type of entity that the WWTP were contracting with and would be directly dependent on the number of trading partners the WWTP needed. If a WWTP wanted to buy credits from farmers, in many cases, it would need to find a large number of them to obtain sufficient credits. The demanding aspects of executing the trade would be multiplied by the large number of trading contracts required.

However, this would vary by region of the country, type of agriculture, and irrigation practices and may be less problematic in some areas than in others.

Executing trading agreements with corporations or government entities may be much simpler. Fewer trading partners (perhaps only one) may be required, and there may be less concern with risk. An excellent example is one of the Rahr Malting trades described in Chapter 2, one of the two trades involving the conversion of erosion-prone agricultural land to native trees and grasses. For this trade, Rahr purchased a perpetual easement from the City of New Ulm, Minnesota, at a one-time cost of $51,200 (the easement is included in its entirety as Appendix C). The core provision of the contract states the following:

> Rahr shall establish and maintain a permanent vegetative cover on the easement area, including any necessary replanting thereof, and other conservation practices. The conservation practices shall include the planting of various native grasses and trees in accordance with the conservation plan for New Ulm properties… Rahr and the City agree that the conservation practices shall be such that they qualify for credits for nonpoint-source trading and wetland conservation act credits.

New Ulm also agreed to various terms that would protect the intended purpose of the conservation actions undertaken by Rahr. This contract is included in its entirety in Appendix C. Note that no quantities of credits appear in the document: the MPCA stipulated how many pounds of phosphorus and CBOD5 credits these land-use changes generated (see Rahr Malting Company section in Chapter 2). While data on the results of this trade alone were not obtained, data are available on the combined performance of the two Rahr agricultural-land conversion trades. During the first five years of the permit, these two trades produced an average of 5.9 kg (13 lb) of phosphorus credits per day. The permit required 8.6 kg (19 lb) per day, and the four trades produced a total of 11.4 kg (25 lb) (Fang and Easter, 2003).

Contracts with Middlemen

A promising alternative to contracting with multiple small sources of credits would be to instead contract with a single entity, which would be responsible for all aspects of delivering the credits. While few such companies offering these types of services currently exist (one example is Environmental Banc and Exchange, Owings Mills, Maryland), in the future they could offer many potential benefits to WWTPs seeking to purchase credits from nonpoint sources. As noted earlier in this chapter, the NASCD has asserted that credit-trading programs can be integrated with state agricultural cost-share programs, to the benefit of both (NASCD, 2003). Regional wetland mitigation banks could possibly fill this role also. A WWTP needing to locate sellers of credits and execute purchases of credits could use the private sector (or one of the above-mentioned public sector agencies) to perform the following functions:

- Find and contract with farmers, businesses, or individual land owners willing to generate and sell credits;
- Secure funding for the BMPs or other improvements, facilities, or equipment needed to generate the credits;
- Obtain all necessary permits;
- Assume responsibility for insuring that all federal, state, and local requirements are met;
- Ensure that BMPs or land-use changes are properly maintained for the life of the contract; and
- Assume risks.

This type of contractual arrangement is already in widespread use by the wastewater community. Contracts with private companies for removal of biosolids from WWTPs for land-application on farmland are very common. The parallel services provided in a land application contract are as follows:

- Find and contract with farmers or land owners willing to accept biosolids application;

- Provide all facilities and equipment needed to transport, store during inclement weather, and land-apply the biosolids;
- Obtain all necessary permits;
- Assume responsibility for insuring that all federal, state, and local requirements are met, including U.S. EPA regulations for use of biosolids; and
- Assume risks.

The contractor assumes risk in several ways. First, the contract invariably requires the contractor to post a performance bond (and sometimes a labor and material bond also) to ensure that farmers and subcontractors are paid by the prime contractor. Second, the contract will contain provisions for failure to perform. These provisions typically state that if the contractor fails to perform, and the WWTP is forced to take other actions to ensure proper disposal of the biosolids, then the contractor is liable for all costs incurred by the WWTP in doing so.

It is critical to note, however, that the WWTP is not freed from the risk of enforcement action by the state or U.S. EPA in the event the contractor fails to comply with state or federal biosolids regulations. Every NPDES permit issued to WWTPs throughout the country contains language similar to the following:

> The permitee shall comply with all existing State and federal laws and regulations that apply to sewage sludge monitoring requirements and utilization practices, and with any regulations promulgated pursuant to Environment Article, Section 9-230 et seq. or to the Clean Water Act, Section 405 (d). The permittee is responsible for ensuring that its sewage sludge is utilized in accordance with a valid sewage sludge utilization permit issued by the Department (NPDES permit for the Piscataway WWTP, Accokeek, Maryland; issued August 1, 2003, by the Maryland Department of the Environment).

Despite having the ultimate responsibility for compliance with this permit requirement, WWTPs have generally found the land-application companies reliable in the performance of these functions and prefer to contract for these services, rather than undertake them themselves. Given the parallels between contracting for water-quality credits versus disposing of biosolids, great potential exists for private companies to function in reliable ways, in both areas.

One final note of caution is for the WWTP to consider including, in its contract with a middleman, a right to assume the middleman's rights and responsibilities under the contract. This right could be critical to the WWTP's ability to terminate the contractor for nonperformance. For example, in the land application of biosolids context, if the land-applier holds the permits to land-apply on the farm fields, it would be difficult or impossible to fire the contractor, because the WWTP would then be left without any permitted fields on which to apply its biosolids.

Taking these factors into consideration, a contract with a private company should include the following:

■ Purpose of the contract;
■ Quantities of credits exchanged;
■ Prices of credits exchanged;
■ Duration of contract;
■ Obligations of the seller*;
■ Obligations of the buyer; and
■ Provisions for violation.

*Obligations of the seller include the following:

■ Obtain performance and/or payment bonds;
■ Find and contract with farmers, businesses, or individual land owners willing to generate and sell credits;
■ Obtain all necessary permits;
■ Ensure the generation of credits;

- Properly maintain BMPs and all facilities needed to generate the credits;

- Allow regular inspections by buyer;

- Comply with all applicable federal, state, and local requirements; and

- Allow the buyer the right to assume all subcontracts upon default by the seller.

The provisions for violation should include a requirement that the contractor assume all liability for costs incurred by the WWTP because of a failure to perform by the contractor, as in the biosolids contracts. This provision may need to be supported by a performance bond, in certain situations.

Trades and National Pollutant Discharge Elimination System Permits

How will water-quality trades be incorporated to NPDES permits? This question is of critical importance to WWTPs, and the manner in which states decide to answer it could have major effects on the viability and attractiveness of water-quality trading. As with trading contracts, this is a new area with no established practices to go by and few actual examples of permits with trading provisions. The Water Quality Trading Policy (U.S. EPA, 2003a) addresses the topic, but only in a general way. Complicating any discussion of the issue is the fact that, as states develop trading programs, the manner in which they choose to address the permitting requirements will vary widely. Hence, at this time, the discussion of trading and permits must remain general.

Trading by NPDES-regulated WWTPs involves regulatory and legal issues; hence, WWTPs interested in pursuing specific trading opportunities should consult with a knowledgeable "sewer" lawyer (i.e. one with CWA expertise). This book only generally explores issues related to water-quality trading and is not intended to provide legal or regulatory advice about any specific trades that may be contemplated by the reader.

U.S. EPA WATER-QUALITY TRADING POLICY PERMIT PROVISIONS

Do water-quality trades have to be reflected in NPDES permits? The answer is an unequivocal yes. The Water Quality Trading Policy states that "Provisions for water quality trading should be aligned with and incorporated into core water-quality programs" (U.S. EPA, 2003a). This statement was included to clarify that all CWA requirements for discharges must be met, including the need for a permit. A discharge that is altered by a trade can still occur only under the terms of an NPDES permit (CWA, 1972; Section 301). The reason that the policy does not explicitly state this is because it is guidance and cannot itself establish regulatory requirements, as it would appear to do if the word "must" were used rather than "should" in the above statement (Batchelor, 2003). In fact, the entire policy is carefully worded to convey what U.S. EPA considers to be the minimum requirements for a trading program, without stating any absolute requirements, unless supported by existing statute or regulation. The policy also states that U.S. EPA "...encourages the inclusion of specific trading provisions in the TMDL itself, in NPDES permits, in watershed plans and the continuing planning process" (U.S. EPA, 2003a).

Wastewater treatment plants should expect (and welcome) having all trades incorporated (or, at least, acknowledged) into their NPDES permits. The challenge is to ensure that state trading programs are practical and achievable and that trades are incorporated into permits in ways that do not unduly harm the viability or attractiveness of trading. The Water Quality Trading Policy is not very prescriptive in its discussion of permits; a few required permit elements are listed, and various possible approaches to others are suggested. Above all, the policy stresses that U.S. EPA supports flexibility in incorporating trades into permits.

Required Elements in a National Pollutant Discharge Elimination System Permit

Several elements must be included in an NPDES permit and/or its accompanying fact sheet. They are

- A description of how the trade was designed, that is, "how baselines and conditions or limits for trading have been established".

- A description of how the trade is consistent with water-quality standards.

- Where the permit or regulations specify methods and procedures (sampling protocols or monitoring frequencies, for example); they should also be applied to the trade where applicable.

- Time periods and pollutant units that are specified in the permit for effluent limits should also be applied to the trade. In other words, the unit of trade (the credit) should be consistent with the mass units used for the effluent limits and the time period during which they are measured and averaged (e.g., weekly, monthly, or annually).

These provisions, by themselves, generally should not raise serious issues or concerns for WWTPs, with the possible exception of the averaging period.

Optional Approaches for Specific Trading Provisions

The Water Quality Trading Policy (U.S. EPA, 2003a) suggests three general methods of including trades in permits.

- General-permit method. A general or watershed permit covering the pollutant of interest is issued collectively to all of the relevant dischargers in the watershed. This permit spells out the individual or collective load allocations and the trading requirements. Examples of this approach are the statewide general permit for nitrogen discharges to Long Island Sound issued by Connecticut and the general nutrient permit issued to the Neuse River Compliance Association in North Carolina. The use of watershed-based permits is discussed later in this chapter.

■ Incorporation-by-reference method. Trading provisions are incorporated to individual NPDES permits by reference to the applicable contract, trading regulations, or requirements of the state's adopted trading program in the General Conditions section of the permit. Numeric limits or performance standards, which drive the trade, are also included. Michigan's regulations on water-quality trading are an excellent example of this approach (Michigan Department of Environmental Quality, 2002).

■ Incorporation-in-full method. Trading provisions are incorporated, in great detail, to the General Conditions sections of individual NPDES permits, through inclusion of the entire contract (as an appendix) or all applicable trading rules and constraints. Numeric limits are also included. The Rahr Malting Company (see Chapter 2) permit is an example of this.

The "general-permit" method would be the least and the "incorporation-in-full" approach would be the most complex and bureaucratically demanding. The "incorporation-by-reference" method would have the benefit of including effluent limits that accommodate trading, while avoiding detailed language on trading requirements. This could provide the flexibility and time needed to work out the terms of trades, locate trading partners, and negotiate contracts. Individual trades would not have to be identified and quantified in the permit, as opposed to the "incorporation-in-full" approach. In addition, if a WWTP were purchasing credits from multiple trading partners, the "incorporation-in-full" method would seem to require that the WWTP know all of its desired trades at the permit renewal application stage. (However, it is possible that a provision could be added that would allow routine changes upon the permitting agency's written approval to be processed administratively as a minor permit modification.)

Inclusion of Numeric Limits

Effluent limits would be included in the permits under all three approaches. For most tradable pollutants, they would be in the form of mass-loading limits. Two different limits could be included: one to apply to the case where no trading occurs and a second limit to apply if trading were to occur. (As noted above, the trading itself would have to be authorized, in some manner, elsewhere in the permit.) The Water Quality Trading Policy lists both "alternate" and "variable" permit limits as possibilities, without describing what is meant by those terms. Further clarification from U.S. EPA indicated that the concepts are closely related, with one difference being that, with alternate limits, the amount of credits to be purchased would have to be decided when the permit is issued. Under variable limits, the amount could vary within a range specified by the permit (Hall, 2003).

Compliance and Enforcement Provisions

The Trading Policy is very clear that trades will be enforceable under the CWA. It states that "mechanisms for determining and ensuring compliance are essential for all trades and trading programs" (U.S. EPA, 2003a).

The main compliance issue contemplated by the Trading Policy, and the only one actually addressed, is the need to ensure that the credits exchanged in a trading program are really generated in the quantities claimed. The policy states that "states and tribes should establish clear enforceable mechanisms consistent with NPDES regulations that ensure legal accountability for the generation of credits that are traded" (U.S. EPA, 2003a).

The inclusion of the phrase "consistent with NPDES regulations", in this sentence, essentially restricts its applicability to WWTPs that are selling credits, because the NPDES regulations would not apply to nonpermitted sources. (Defaults by nonpoint-source sellers could be addressed in the state's trading program legislation or regulations.) For a WWTP, then, a failure to actually generate the full amount of the credits sold would constitute a violation of its NPDES permit and would expose it to the full range of penalties under the CWA. In addition, the WWTP using those credits would also be affected.

> In the event of default by another source generating credits, an NPDES permittee using those credits is responsible for complying with the effluent limitations that would apply if the trade had not occurred (U.S. EPA, 2003a).

While this initially would not be a noncompliance, it would become one if the WWTP were not able to quickly comply, either by reducing its own discharge to within the "no-trade" effluent limit or by acquiring credits from another source. This is perhaps the biggest regulatory concern for WWTPs regarding trading.

In the situation where the credits are supplied by nonpoint-source reductions, the policy recognizes that nonpoint-source-control measures can be overwhelmed by extreme natural events.

> [U.S.] EPA recommends that states and tribes consider including provisions to address situations where nonpointsource controls and management practices that are implemented to generate credits fail due to extreme weather conditions or other circumstances that are beyond the control of the source (U.S. EPA, 2003a).

This statement is general enough (i.e., "other circumstances that are beyond the control of the source...") that it should alleviate some of the concern by WWTPs about defaults by nonpoint-source credit suppliers.

In addition, where a trade with a nonpoint source produces benefits beyond the pollutant of interest, the WWTP should make the regulatory agency aware of this fact and should try to make it a major consideration in the event that there are infrequent noncompliances by the nonpoint source in generating the credits.

Overall, the Trading Policy requirements create compliance and enforcement risks for WWTPs that go beyond the normal ones associated with the operating of their own facilities in compliance with an NPDES permit. This comes from having to rely on the actions of other parties to meet permit requirements. Possible methods for minimizing this risk are discussed in the conclusion of this chapter.

Public Opportunity for Comment and Avoidance of Permit Modifications

Federal regulations require that the public be afforded an opportunity to review, comment on, and request a public hearing for all NPDES permits (40 CFR 124 [Procedures for Decisionmaking, 2004]). This creates the opportunity for the public to comment on a proposed trade if the draft permit contains authorization and provisions for it. To provide the public this opportunity, the Trading Policy states the following:

> NPDES permits and fact sheets should describe how baselines and conditions or limits for trading have been established and how they are consistent with water quality standards (U.S. EPA, 2003a).

If a proposed permit contains all of these elements—authorization to trade, provisions for trading, derivation of baselines, and demonstration of consistency with water-quality standards——and the public has been given the opportunity to review and comment on it, then the permit, once issued, would not have to be reopened and modified to incorporate individual trades. The Trading Policy states

> [U.S.] EPA does not expect that an NPDES permit would need to be modified to incorporate an individual trade, if that permit contains authorization and provisions for trading to occur and the public was given notice and an opportunity to comment and/or attend a public hearing at the time the permit was issued (U.S. EPA, 2003a).

This avoids the potentially onerous possibility that permits would have to be frequently modified to incorporate specific trades, a time-consuming process that would require the same opportunity for public comment that the initial issuance of the permit did.

On the other hand, actions that constitute modifications to a TMDL, such as altering the wasteload or load allocations, could result in a requirement to reissue NPDES permits.

In general, it would be advisable for the WWTP to include detailed descriptions of its trading-related aspirations, plans, ongoing activities, etc., in NPDES permit applications.

WATERSHED-BASED DISCHARGE PERMITS

The U.S. EPA has been working to improve the NPDES permitting process so that it can facilitate and promote watershed-based actions and more flexible, efficient, and cost-effective ways of achieving watershed goals, including water-quality trading. In a December 3, 2002, memo, G. Tracy Mehan, U.S. EPA Assistant Administrator for Water, reiterated U.S. EPA's commitment to a watershed approach to water-quality management and directed that a number of program activities designed to advance watershed management be undertaken (U.S. EPA, 2002). One of these directives was to "accelerate efforts to develop and issue NPDES permits on a watershed basis". This memo was followed by a policy statement on watershed-based permitting (U.S. EPA, 2003b).

The essence of watershed-based permitting is contained, in one brief passage, in the policy statement.

> Watershed-based permitting is a process that ultimately produces NPDES permits that are issued to point sources on a geographic or watershed basis. In establishing point-source controls in a watershed-based permit, the permitting authority may focus on watershed goals, and consider multiple pollutant sources and stressors, including the level of nonpoint-source control that is practicable (U.S. EPA, 2003b).

The policy states that U.S. EPA believes that watershed permitting can do the following:

- Lead to more environmentally effective results;
- Emphasize measuring the effectiveness of targeted actions on improvements in water quality;
- Provide greater opportunities for trading and other market-based approaches;
- Reduce the cost of improving the quality of the nation's waters;
- Foster more effective implementation of watershed plans, including TMDLs; and

■ Realize other ancillary benefits, beyond those that have been achieved under the CWA (e.g., facilitate program integration, including integration of CWA and Safe Drinking Water Act [1974] programs).

The policy statement was followed by the publication of detailed guidance on watershed-based permitting in December, 2003 (U.S. EPA, 2003c). The guidance describes four watershed permitting approaches.

(1) Watershed-based general permit—common sources. This type of permit would be issued to a category of dischargers in a watershed, such as all WWTPs. The only difference between this and existing general permits is that the existing permits are not issued on a watershed basis.

(2) Watershed-based general permit—collective sources. This type of permit would be identical to the one above but would cover all point-source dischargers in the watershed, regardless of category, or could cover a subset of the categories (e.g., publicly owned treatment works [POTWs] plus municipal stormwater).

(3) Watershed-based individual permit—multiple permittees. This type of permit would allow a group of dischargers to be covered under a single permit. The Neuse River Compliance Association permit, described in the Neuse River Compliance Association General Permit section below, is an example of this type of permit, as is the Connecticut nitrogen permit described in Chapter 2.

Neuse River Compliance Association General Permit

The Neuse River Compliance Association and its members were issued an NPDES permit (NCC000001) effective January 1, 2003, for total nitrogen discharges. This watershed permit took the "co-permittee approach", described in the watershed-based permitting guidance. The nitrogen allocations are based on a TMDL, as measured at the estuary, which extends from approximately Streets Ferry to the Pamlico Sound. The association was assigned an allocation, as a whole; however, in addition, the members were assigned individual allocations. The co-permittee total nitrogen allocations are also included in the individual NPDES permits of each of the members. A co-permittee member's estuary total nitrogen load is equivalent to its discharged total nitrogen load, multiplied by the applicable transport factor (ranging from 0.1 to 1.0). If the association exceeds its collective allocation, enforcement actions could be taken against the association and any member that exceeded its individual allocation. In addition, if the association cap is exceeded, offset payments must be made to the Wetland Restoration Fund, at a rate of $11 per kilogram (pound) (Holt, 2004).

(4) Integrated municipal NPDES permit. In this approach, all of the possible permits for a given municipality (e.g., wastewater discharge, stormwater, and combined sewer overflow) would be combined in a single permit. Attempts to develop this type of permit by the Louisville and Jefferson County Metropolitan Sewer District in Louisville, Kentucky, and Clean Water Services in Oregon are described below.

Integrated Municipal Permits

Louisville and Jefferson County Metropolitan Sewer District

The Louisville and Jefferson County Metropolitan Sewer District (MSD), Louisville, Kentucky, has been exploring the watershed permit concept for several years. With encouragement from U.S. EPA and the Kentucky Division of Water (KY DOW), MSD evaluated its monitoring programs related to its various NPDES permits and reorganized these programs to streamline monitoring activities. Employees were cross-trained, and assignments were adjusted to optimize efforts needed to comply with permits for MS4, combined sewer overflow, and sanitary sewer overflow programs.

The MSD proceeded with a pilot consolidated-permit application for Beargrass Creek to cover the three programs and execute a

TMDL to respond to findings of impairment by low dissolved oxygen and fecal bacteria. The permit would allow MSD to determine the optimal combination of actions, under the three programs, to achieve TMDL allocations. The agency has established 21 monitoring stations in Beargrass Creek and has planned a series of discrete sampling programs to define the pollutant contribution of various land uses.

The MSD submitted its permit application to KY DOW in July 2003 but had not received comments as of April 2004. Budget constraints in the state agency and loss of some key staff are probably the cause of the delay. However, MSD remains optimistic that a consolidated permit will eventually be issued (Grace, 2004).

Clean Water Services

Clean Water Services (CWS), formerly the Unified Sewerage Agency of Washington County (Oregon), operates four WWTPs that discharge to the Tualatin River watershed, which Oregon Department of Environmental Quality (DEQ) determined was impaired. A TMDL was adopted in 1988 to cover phosphorus and ammonia; a revised TMDL in 2001 established wasteload allocations for temperature to protect salmonid rearing.

The DEQ issued a watershed permit to CWS in February 2004 to cover the four treatment plants and stormwater discharges from the plant properties, the MS4, and the industrial pretreatment program. The permit provides for water-quality trading in two areas: oxygen-demanding discharges from two of the WWTPs and temperature.

Clean Water Services is permitted to trade between carbonaceous biochemical oxygen demand, which is exerted mostly by biodegradable organic matter, and nitrogenous oxygen demand, which is mostly ammonia. Clean Water Services is also able to trade between the two WWTPs, in accordance with an oxygen-depletion model for the receiving stream. The permit does not provide for trading between stormwater and WWTP loadings.

The temperature-trading program was devised to achieve warm-weather temperature standards without expending the capital and operating funds to install refrigeration systems at the treatment plants. Clean Water Services must provide DEQ with a thermal-load-credit trading plan, which will be based on three methods for meeting stream temperature standards: effluent reuse for irrigation of nonfood crops, offsetting water withdrawals from the stream by farmers; flow augmentation with water controlled by CWS in two reservoirs in the watershed; and riparian shading in the upper portion of the watershed (Oregon Department of Environmental Quality, 2004).

A common theme of these four approaches is that they focus on the watershed goals and provide flexibility for achieving those goals in the most efficient way. In the integrated permit approach, for example, trade-offs can be made between stormwater control measures and WWTP effluent limits to achieve water-quality standards in the most cost-effective way. The first three types of permits could facilitate water-quality trading, in a variety of ways, and a WWTP interested in trading should carefully evaluate the potential benefits of a watershed-based permit.

WASTEWATER TREATMENT PLANT GOALS FOR NATIONAL POLLUTANT DISCHARGE ELIMINATION SYSTEM PERMITS

The prospect that water-quality trades will be incorporated to NPDES permits has raised a number of concerns among WWTPs. A few of these concerns are valid; however, many are misplaced. If states and WWTPs are careful, incorporating trades into permits will not create undue risks or difficulties for WWTPs. That said, the first rule of NPDES permit negotiations should always be kept in mind: "Never, ever accept a requirement in the permit that you are not sure you can meet" (Calamita, 2000). If done properly, trading programs will not introduce any permit elements that would violate this commonsense advice.

Ideally, permit requirements should be reliably achievable, practical, workable, as simple as possible, and they should not create undue compliance and enforcement risk. Achieving these characteristics in trading provisions means first insuring that the trading program requirements adopted by the state, in the creation of the trading program, do not prevent this outcome. Sound and workable trading program rules should result in permit requirements with the same characteristics.

A prime example is the need for the adoption of appropriate units to define credits and establish averaging periods to measure compliance (Batchelor, 2003). If a TMDL establishes allocations for phosphorus based on average annual loads, for example, then the incorporation of monthly and weekly phosphorus load limits to discharge permits would probably be a fatal blow to any prospect for trading because of the difficulty and expense of quantifying and tracking phosphorus loads on a weekly basis, by both the user and provider, especially if the provider were a nonpoint-source provider. In addition, WWTP's generally perceive weekly limits to be statistically riskier than monthly limits, and monthly ones riskier than annual ones (and annual ones perhaps riskier than rolling annual average ones). As a result, the shorter the required averaging period, the more conservative the design and operating strategies chosen by the WWTP and the higher the costs. The averaging period specified in the permit should be the same as the analytical framework that established the allocations. The permitting authority should only require a shorter averaging period if there is a documented water-quality reason to do so.

Justification for Annual Permit Limits for Nitrogen and Phosphorus in the Chesapeake Bay Watershed

In the development of a permitting strategy for nutrients for the Chesapeake Bay, U.S. EPA Region III and the Chesapeake Bay Program Office requested that the U.S. EPA Office of Water address the question of whether annual permit limits for nitrogen and phosphorus could be contemplated instead of conventional monthly or weekly limits. In a March 3, 2004, memo (U.S. EPA, 2004), Jim Hanlon, Director of the Office of Wastewater Management, responded as follows:

I concur that permit limits expressed as an annual limit are appropriate, and it is reasonable, in this case, to conclude that it is "impracticable" to express effluent limitations as daily maximums, weekly average, or monthly average effluent limitations (U.S. EPA, 2004).

The memo justifies this position by finding that nutrient limits are different from parameters, such as toxic pollutants, because

- The exposure period of concern is very long,
- The area of concern is far-field rather than near-field, and
- The average pollutant load rather than the maximum load is of concern.

While the memo states that these findings may not apply to situations smaller, in scale, than the Chesapeake Bay, it also points out that the nutrient dynamics of the bay may not be unique. If nutrient management issues in other watersheds share the three characteristics listed above, then annual limits may be appropriate. Their use, however, must be supported by "robust data and modeling" and appropriate safeguards must be used to protect all other applicable water-quality standards. (Note: The Chesapeake Bay water-quality model was used to assess annual versus monthly load limits for both nitrogen and phosphorus [i.e., varying versus constant monthly loads with the same annual load]. For both nutrients, the modeled water-quality results of the two scenarios were indistinguishable.)

Conclusion

The compliance issue that should most concern WWTPs is the potential default of a provider of credits. As described above, the using WWTP would then be required to comply with the permit limits that would be in effect in the absence of trades. If it were unable to do so, it would be in noncompliance with its permit and subject to the full range of CWA legal sanctions. Wastewater treatment plants are understandably wary of being held liable for the actions of a trading partner. However, it is not unreasonable to argue that pollutant reductions achieved through enforceable NPDES permit limits cannot simply be traded away for unenforceable nonpoint-source reductions with no CWA recourse by the state, in the event of a default.

In crafting trading programs, states and WWTPs must find ways to deal with this issue without unduly harming the attractiveness and viability of trading. There are several possible methods that can be explored with the state in the development of the trading program (Batchelor, 2003; Calamita, 2003; Hall, 2003). They include the following:

- ■ The choice of stable, reputable trading partners.

- ■ The inclusion of provisions in contracts to protect the buyer in the event of default by the seller. Such provisions could include performance bonds, payment bonds, insurance requirements, and the holding of the seller liable for any costs incurred by the buyer because of a failure to perform by the seller (see the discussion of contracts earlier in this chapter for more detail).

- ■ The use of trading ratios, state-approved BMPs, and state-stipulated BMP performances. If these methods for quantifying nonpoint-source trades are adopted, then the WWTP using the credits needs only to try to ensure that the provider is maintaining the BMPs in accordance with state and contractual requirements. If so, the credits are real and valid by definition.

- ■ In individual applications, a trading arrangement could be embodied in both a permit and state administrative order. The order could establish stipulated penalties, at a reasonable level, for any noncompliance. The stipulated penalties would provide a firm (yet reasonable) basis to seek indemnification from the WWTP's trading partner. Such a long-term state administrative order could take much of the risk out of nonpoint-source trading. The stipulated penalties could be modest if the trade produces indirect pollutant reductions or benefits beyond the specific parameter that is the subject of the trade, especially if a municipal WWTP (as opposed to an industrial WWTP) is involved. While potentially not an absolute bar against U.S. EPA or third-party enforcement, state orders to address any trading issues may warrant consideration.

■ The existence of a "reconciliation and truing-up" period at the end of the averaging period, during which time credit shortfalls could be rectified by acquiring credits from one of the available backup mechanisms. The Michigan Rules provide for such a period (see Appendix A) (Michigan Department of Environmental Quality, 2002). If construction is needed, a reasonable compliance schedule before the "non-trade" limits become effective should be included.

■ The availability of "trading banks" maintained by the state as an emergency or backup source of credits in the event the WWTP is unable to meet its permit requirements. This option is available to both the Tar-Pamlico and Neuse Associations and is being contemplated by Maryland for its nutrient trading program for the Chesapeake Bay (Rhoderick, 2003).

Of these five, only the last two would be foolproof methods for insuring that a WWTP would not face a noncompliance because of a default by its credit supplier. It is strongly recommended that WWTPs work closely with the state during the development of the trading program to try to ensure that a reliable backup source of credits would always be available and a reconciliation period that would give the WWTP time to replace credits before going into noncompliance with the permit. If these elements were in place, the compliance and enforcement worries by the wastewater treatment community would be minimized.

References

Batchelor, D., U.S. Environmental Protection Agency, Washington, D.C. (2003) Personal communication.

Calamita, P., AquaLaw PLC Water and Wastewater Solutions, Richmond, Virginia (2000) Personal communication.

Calamita, P., AquaLaw PLC Water and Wastewater Solutions, Richmond, Virginia (2003) Personal communication.

Carter, D. L. (2002) Proposed Best Management Practices to be Applied in the Lower Boise River Effluent Trading Demonstration Project. Unpublished report; http://www.envtn.org/docs/carter2002bmps.PDF (accessed April 17, 2004).

Clean Water Act (1972) U.S. Code, Section 1251–1387, Title 33.

Dillaha, T. A. (1998a) Influence of the Hydrologic Cycle on Nonpoint Source Pollution. Unpublished paper prepared for the Chesapeake Bay Program Scientific and Technical Advisory Committee, Chesapeake Research Consortium, Inc.: Edgewater, Maryland.

Dillaha, T. A. (1998b) Role of BMPs in Reducing NPS Pollution. Unpublished paper prepared for the Chesapeake Bay Program Scientific and Technical Advisory Committee, Chesapeake Research Consortium, Inc.: Edgewater, Maryland.

Faeth, P. (2000) *Fertile Ground—Nutrient Trading's Potential to Cost-Effectively Improve Water Quality*; World Resources Institute: Washington, D.C.

Fang, F.; Easter, K. W. (2003) Pollution Trading to Offset New Pollutant Loadings–A Case Study in the Minnesota River Basin. Paper presented at the American Agricultural Economics Association Annual Meeting, Montreal, Canada, July, 2003; http://www.envtn.org/docs/MN_case_Fang.pdf (accessed March 19, 2004).

Grace, P., Water Quality/Quantity Administrator, Louisville and Jefferson County Metropolitan Sewer District, Louisville, Kentucky (2004) Personal communication.

Gray, M. (2003) Environmental Credit Trading: A Next-Generation Incentive for Conservation on Agricultural Lands. Presented at the National Forum on Water Quality Trading, Chicago, Illinois, July.

Hall, L., U.S. Environmental Protection Agency, Washington, D.C. (2003 and 2004) Personal communication.

Holt, L., Town of Cary, North Carolina (2004) Personal communication.

Idaho Soil Conservation Commission (2002) Best Management Practice (BMP) List for the Lower Boise River Pollution Trading Program. Unpublished report; http://www.envtn.org/docs/boise_bmp_manual_DRAFT.doc (accessed April 17, 2004).

Kerr, R. L.; Anderson, S. J.; Jacksch, J. (2000) Crosscutting Analysis of Trading Programs—Case Studies in Air, Water and Wetlands Mitigation Trading Systems. Research paper prepared for the National Academy of Public Administration, Washington, D.C. http://www.napawash.org/pc_economy_environment/epafile06.pdf (accessed April 10, 2004).

Kieser, M. S. (2000) *Phosphorus Credit Trading in the Kalamazoo River Basin: Forging Nontraditional Partnerships*; Water Environment Research Foundation: Alexandria, Virginia.

Kieser, M. S. (2003) Environmental Multiple Markets—A White Paper on the Concept (Draft). http://www.envtn.org/docs/emm_es.PDF (accessed April 9, 2004).

King, D. M.; Kuch, P. J. (2003) Will Nutrient Credit Trading Ever Work? An Assessment of Supply and Demand Problems and Institutional Obstacles. 33 *Environ. Law Reporter News*, Washington, D.C., 10352–10368, http://www.envtn.org/ELR_trading_article.PDF (accessed January 21, 2005).

Lanyon, L. E. (1998) Estimating Non-Point Nutrient Loads from Agriculture for Pollution Permit Trading. Unpublished paper prepared for the Chesapeake Bay Program Scientific and Technical Advisory Committee, Chesapeake Research Consortium, Inc.: Edgewater, Maryland.

Michigan Commission of Agriculture (1977a) *Generally Accepted Agricultural and Management Practices for Nutrient Utilization*; Michigan Commission of Agriculture: Lansing, Michigan.

Michigan Commission of Agriculture (1977b) *Generally Accepted Agricultural and Management Practices for Manure Management and Utilization*; Michigan Commission of Agriculture: Lansing, Michigan.

Michigan Commission of Agriculture (1977c) *Generally Accepted Agricultural and Management Practices for Pesticide Utilization and Pest Control*; Michigan Commission of Agriculture: Lansing, Michigan.

Michigan Department of Environmental Quality (2002) Rule Part 30: Water Quality Trading (effective November 22, 2002).

Michigan Department of Environmental Quality: Surface Water Quality Division: Lansing, Michigan; http://www.state.mi.us/orr/emi/arcrules.asp?type=Numeric&id=1999&subId=1999%2D036+EQ&subCat=Admincode (accessed April 16, 2004).

Minnesota Pollution Control Agency (1997) Final Issuance NPDES/SDS Permit MN 0031917; Rahr Malting Co: St. Paul, Minnesota.

National Association of Soil Conservation Districts (2003) Water Quality Trading—Nonpoint Credit Bank Model. Unpublished report; http://www.envtn.org/resources/TradingBankModel-Paper.doc (accessed April 6, 2004).

National Permit Program (2004) Code of Federal Regulations, Part 129, Title 40.

North Carolina Department of Natural Resources (1994) *Non-Point-Source Management Program: Tar-Pamlico Nutrient Strategy*; North Carolina Division of Water Quality: Raleigh, North Carolina; http://h2o.enr.state.nc.us/nps/tarpam.htm (accessed April 17, 2004).

Oregon Department of Environmental Quality (2004) National Pollutant Discharge Elimination System Watershed-Based Waste Discharge Permit Issued to Clean Water Services, Portland, Oregon; http://www.deq.state.or.us/wq/wqpermit/indvpermit-docs.htm (accessed April 17, 2004).

Paulson, C.; Vlier, J.; Fowler, A.; Sandquist, R.; Bacon, E. (2000) *Phosphorus Credit Trading in the Cherry Creek Basin: An Innovative Approach to Achieving Water Quality Benefits*; Water Environment Research Foundation: Alexandria, Virginia.

Rhoderick, J., Maryland Department of Agriculture, Annapolis, Maryland (2003) Personal communication.

Ross and Associates (2000) Lower Boise River Effluent Trading Demonstration Project: Summary of Participant Recommendations for a Trading Framework. Report prepared for the Idaho Division of Environmental Quality. http://www.deq.state.id.us/water/tmdls/lowerboise_effluent/lowerboiseriver_effluent.htm(accessed April 17, 2004).

Safe Water Drinking Act (1974) Public Law 104-182; Code of Federal Regulations, Title 42.

Schary, C., U.S. Environmental Protection Agency, Washington, D.C. (2000) Personal communication.

Stacey, P., Connecticut Department of Environmental Protection, Hartford, Connecticut (2004) Personal communication.

U.S. Environmental Protection Agency (2001) *Chesapeake Bay Program Nutrient Trading Fundamental Principles and Guidelines;* EPA-903/B-01-001; Chesapeake Bay Program, Nutrient Trading Negotiation Team: Rockville, Maryland; http://www.chesapeakebay.net/trading.htm (accessed June 20, 2004).

U.S. Environmental Protection Agency (2002) Memo from G. Tracy Mehan, III. Committing EPA's Water Program to Advancing the Watershed Approach. http://cfpub.epa.gov/npdes/wqbasedpermitting/wspermitting.cfm. (accessed April 8, 2004).

U.S. Environmental Protection Agency (2003a) Water Quality Trading Policy. Unpublished guidance; http://www.epa.gov/owow/ watershed/trading/finalpolicy2003.html (accessed June 20, 2004).

U.S. Environmental Protection Agency (2003b) Watershed-Based NPDES Permitting Policy Statement. http://cfpub.epa.gov/npdes/ wqbasedpermitting/wspermitting.cfm (accessed April 8, 2004).

U.S. Environmental Protection Agency (2003c) Watershed-Based National Pollutant Discharge Elimination System (NPDES) Permitting Implementation Guidance, EPA-833/B-03-004. http://cfpub.epa.gov/npdes/wqbasedpermitting/wspermitting.cfm (accessed April 8, 2004).

U.S. Environmental Protection Agency (2004) Annual Permit Limits for Nitrogen and Phosphorus to Protect Chesapeake Bay and its Tidal Tributaries from Excess Nutrient Loading under the National Pollution Discharge Elimination System. Memo from J. A. Hanlon, Office of Wastewater Management to J. Capacasa, EPA Region 3 and R. Hamner, Chesapeake Bay Program Office, March 3; Washington, D.C.

Woodward, R. T. (2003) Lessons about Effluent Trading from a Single Trade. Unpublished paper; http://ageco.tamu.edu/faculty/ woodward/paps/CaseStudy.pdf (accessed April 17, 2004).

CHAPTER SIX

Science, Data, and Analytical Needs

Introduction

The mission of achieving and maintaining desired aquatic environmental conditions requires the means to (1) define desired and achievable environmental conditions specific to a watershed or water body; (2) reliably measure or predict current environmental conditions; (3) differentiate between the contribution of natural processes and the activities of society to those environmental conditions; and (4) predict the extent and timeframe of improvements to the aquatic environment that would accrue from a pollution-control strategy. The first element is predicated on a shared and agreed-to vision by the stakeholders. The remaining three elements depend on data and models that cannot be defined with certainty. The definition of environmental conditions is affected by the design and execution of the monitoring program. Differentiating contributions from natural pollutant sources and allocating the human effect among the various activities in a watershed depends on models of physical and biochemical processes. Predicting the improvements that would result from various pollution-control strategies also requires modeling. Further, pollution-control strategies will invariably consist of political and scientific components.

The uncertainty that pervades programs to protect the aquatic environment is not unique to this endeavor; it affects virtually all natural processes and human activity. In many of our routine tasks (i.e., working at our daily jobs), uncertainty exists. In the daily commute to work, we might be very certain that the building will be there when we arrive, but the drive to the workplace is subject to traffic and weather, which interrelate in ways that we understand better as we gather and digest empirical data from each day's trip. Once in the workplace, uncertainty increases as our agenda is affected by human and environmental interventions. As we gain experience in our jobs, we develop monitoring programs to alert us to impending disturbances and models to guide us in dealing with the disturbances, while also completing the work on the agenda.

This workday analogy (and many others that could be extracted from human endeavors) is more closely related to the mission of managing the aquatic environment than might first be apparent. A monitoring program may rest on a certainty that water will flow downhill, but all other knowledge of the status of the aquatic environment must be measured, yielding empirical data. These data are affected by numerous factors, many of them interrelated, such as time of day and time of year, weather, changes in land use, sampling procedures, and analytical methods. The list could go on for many pages.

Identifying pollution sources and allocating cleanup responsibility requires models that have plausible connection to actual natural processes and a good body of data for calibration. These models are constrained by the limits of our understanding of natural processes and by practical limits on the detail that we can accommodate within funding and time constraints.

Success in the mission of managing the aquatic environment depends on plenty of good data, practical models that relate to natural processes in plausible ways, and the attention of experienced human participants. All of these requirements entail funding and time demands. Looking back at the workday analogy, we assemble and analyze data, develop models gradually over time to guide our activities, and make many decisions during each day as to how to do the things that must be done, while dealing with ongoing uncertainty. We decide how to spend our time, which also represents funding, to achieve a reasonable balance among competing demands.

Two Aspects of Uncertainty: Science and Humans

When we consider uncertainty in water-quality-management programs, we generally think of data and modeling. Here, we will call data and modeling "science." We realize that the uncertainty entailed in the science of water-quality management can be addressed and defined by statistical analysis. This "scientific" approach to uncertainty offers a degree of comfort: we can establish boundaries to our uncertainty that can generally be honored by obtaining more data or by more extensive calibration of models (which means using more data). The first part of this chapter will identify the uncertainty that can be accommodated by statistical analysis, along with some comments on appropriate statistical treatment.

The human element is more difficult to factor into our thinking because it cannot be modeled as a natural system. It is harder to predict than natural-system uncertainty, but experience in the objectives that the various parties bring to the water-quality-management process will improve one's ability to deal with this aspect of uncertainty. The human element will be explored in the second part of this chapter.

Getting into the Spirit: Collaborate, Coordinate, and Communicate

Water-quality-management programs are founded on two elements: (1) defining the status of an aquatic system under conditions prevailing at the time of observation and (2) predicting the status of the system under hypothetical conditions. These hypothetical conditions might not be future or speculative conditions; they might represent an actual condition that has not yet been measured. The status of a system is determined through monitoring programs, and system status under hypothetical conditions is predicted by modeling. Both of these efforts depend on an appropriate body of data that meet the quality objectives for their use.

Water-quality data exist in two general forms: evaluated data and monitored data (NRC, 2001). Evaluated data sets are generally older and smaller than more recent data derived from defined monitoring programs, or they may be inferred from regulatory actions. A record of beach closures, for example, might be inferred to represent a fecal coliform criterion exceedance. Monitored data, on the other hand, have been derived from programs that are defined by information that allows the user to decide whether the data meet the quality objectives for their use. The data about the data, often referred to as metadata, are as important as the data that define water-quality status. Evaluated data might serve to indicate a trend or allow a general view of water quality in the past, but they entail uncertainty that cannot be determined because the nature of the monitoring program that produced them is unknown.

Surrogate criteria and indicators are often used in the water-quality-management field. For example, nutrient control is often used to achieve dissolved-oxygen standards in receiving water bodies. Nitrogen and phosphorus are linked to dissolved oxygen through models that predict algae growth and the oxygen demand that the algae exert through nighttime respiration and as they die and decompose. Wang and Kanehl (2003) note that stream macroinvertebrates are often used as surrogate indicators of general stream health. Also, most models that are used to predict sediment loads from storm runoff use land cover and development as surrogate criteria that incorporate the many variables that determine the tendency of soils to erode.

A successful water-quality-monitoring program to define the status of a given water body under particular circumstances would consist of several elements

- ■ Program design;
- ■ Field sampling or observation;
- ■ Laboratory analysis, if applicable;
- ■ Data management; and
- ■ Data interpretation.

Many valid monitoring programs could be conceived to include these five elements, but the resulting data might or might not be useful for application to other water bodies. While the data might be sufficient to allow for site-specific, water-quality decisionmaking, they might have limited usefulness on other sites. Thus, each situation could require an extensive—and costly—monitoring program. For example, modeling requires large data sets for calibration purposes. If data were obtained by monitoring programs designed and executed within a framework with national scope, they would likely be useful in similar situations in other locations.

Certain members of the water-quality-management community recognized the potential usefulness of a standard approach to monitoring by forming the Intergovernmental Task Force on Monitoring Water Quality (ITFM) in 1992. Headed by the U.S. Geological Survey and the U.S. Environmental Protection Agency (U.S. EPA), the task force included federal, state, tribal, and interstate agencies with interests in and regulatory responsibility for water quality. The ITFM recognized that the emphasis in the nation's effort to attain the goals of the 1972 Clean Water Act (CWA) would necessarily have to change from point- to nonpoint-pollution sources, and that this shift in emphasis would be facilitated by broad adoption of water-quality-monitoring programs. The task force issued a final report in 1995 that established a general framework for monitoring programs (ITFM, 1995).

In 1997, the ITFM was reformed as the National Water Quality Monitoring Council (NWQMC) to continue the work of developing a comprehensive framework for water-quality-monitoring program design and execution. The NWQMC convened three national monitoring conferences (in 1998, 2000, and 2002) and has developed a proposed framework consisting of six tasks that must occur to produce data that would support water-quality decisionmaking and regulation

(1) Develop monitoring objectives,

(2) Design monitoring program,

(3) Collect field and laboratory data,

(4) Compile and manage data,

(5) Assess and interpret data, and

(6) Convey results and findings.

These six tasks must be carried out in an atmosphere that fosters collaboration, coordination, and communication. This framework is illustrated in Figure 6.1.

The circular format chosen by NWQMC illustrates the view of the council that a meaningful water-quality-monitoring program should undergo continuous assessment and improvement. This spirit of continuous improvement is important for all parties involved in

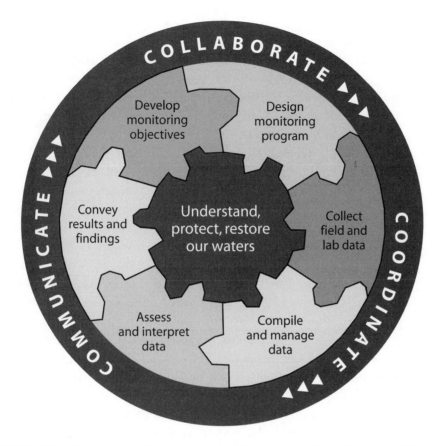

FIGURE 6.1 Framework for monitoring (courtesy of the National Water Quality Monitoring Council).

the mission of water-quality management to understand and embrace. It pervades the recommendations of the National Research Council (NRC, 2001) and the General Accounting Office (GAO, 2000). The NWQMC webpage (http://water.usgs.gov/wicp/acwi/monitoring) offers a number of links to success stories, including a description of the Connecticut River Watch Program (CRWP). This program, which is overseen by an advisory committee formed by federal, state, and local agencies and organizations, has involved some 200 volunteers assisting the Connecticut Department of Environmental Protection (DEP) in sampling and documenting the condition of the Mattabesset River. Efforts by CRWP, Connecticut DEP, and the Middlesex County Soil and Water Conservation District have resulted in a watershed management plan that is supported by local governments, business interests, residents, and environmental groups. This and other success stories illustrate the power of collaboration in achieving environmental improvements that require concerted action by many parties.

Although developing total maximum daily loads (TMDLs) and allocating pollution-control responsibility can be complex, many practitioners in the field report success in achieving stakeholder acceptance through proactive, transparent, and collaborative processes. Many of the presenters at the Water Environment Federation®/Association of State and Interstate Water Pollution Control Administrators TMDL 2001 Conference in St. Louis, Missouri, including Brown (2001), Hansen (2001), Hauck and Vargas (2001), Lee (2001), Stiles (2001), and Stober et al. (2001), confirmed the positive results of the collaborative approach.

In his editorial introducing a special TMDL issue of the American Society of Civil Engineers (ASCE) *Journal of Water Resources Planning and Management*, Reckhow (2003) addressed uncertainty in water-quality modeling and endorsed the adaptive approach to watershed and water-quality management. His endorsement is based on the uncertainty that pervades current modeling practice and the data collection that supports it. He suggests that the stakeholder community can work together to provide motivation and guidance to decision-makers as to how uncertainty should be factored into TMDLs.

These examples provide heartening evidence that the regulatory atmosphere may be evolving away from a "command-and-control" approach toward cooperation and collaborative problem solving. While this evolution might be farther advanced in some states than in others, wastewater treatment plant (WWTP) management, in all states, should adopt a positive attitude and enter fully into the process. Gradually, the methodical involvement of stakeholders in watershed-based strategies for water-quality management, along with transparent model development, testing, and application, should prevail. The alternative—litigation, with dueling experts and data—would be inordinately costly, in both time and funds, when compared to a growing body of successful collaborative efforts.

A WWTP should not join the process with the idea of delaying or derailing watershed-based, water-quality management. Undue delays, particularly when they are induced by the WWTP, may encourage either the regulatory agency to adopt command-and-control tactics or environmental advocacy groups to initiate litigation. Under either circumstance, the WWTP would probably present the easiest target for control, through the National Pollutant Discharge Elimination System (NPDES) program. This chapter will provide some insights to data quality and modeling practices; however, these observations should be considered in the context of collaboration, communication, and coordination. The mission of managing water quality on a watershed basis is fraught with uncertainty, and such uncertainty could always be cited as reason for inaction. A WWTP that adopts the approach of challenging and criticizing will probably find itself marginalized and unable to affect the outcome of water-quality-strategy development.

The Water Environment Research Foundation (WERF) has issued a series of reports under the title *Navigating the TMDL Process*. One of these, Evaluation and Improvements (WERF, 2003a), would be a particularly good choice for the WWTP manager who is interested in understanding the range of issues pertaining to the TMDL process and recommendations of improvements. A broad understanding of the process will be important to allow the WWTP manager to participate, in meaningful ways, in the process.

Uncertainty in the Science: Data and Modeling

DATA ISSUES

Errors will be introduced to water-quality data from a number of sources

■ Biases in monitoring program;

- ■ Sampling errors, at the sampling point and during transport; and

- ■ Laboratory errors, including method, instrument, and human errors.

Biases can often be detected by plotting data, an approach described by Berthouex and Brown (2002), and the monitoring program can be modified to remove them. Variabilities from the other two causes are likely to be random; they can be accommodated by statistical methods. Natural anomalies may also affect data sets. For example, seasonal lake turnover may cause samples taken at a single location to vary significantly during the course of a year. This kind of effect would be similar to a bias and should be discernible by plotting the data with respect to season.

We turn to statistical analysis to account for these inevitable errors, posing a pair of hypotheses. The null hypothesis is that the water body is not impaired. The alternative hypothesis is that the water body is impaired. Obviously, this pair of hypotheses covers all possibilities. Two types of error are defined.

- ■ Type I error. A water body that is not impaired will be erroneously identified as impaired.

- ■ Type II error. An impaired body of water will not be identified as impaired.

A data set can be tested by a number of statistical methods. Perhaps the simplest test is the raw-score test, which sorts the data into two categories: (1) less than a numeric water-quality criterion or (2) greater than the criterion. A tolerable exceedance rate is identified, and, if the "greater-than" group comprises a rate higher than the tolerable exceedance rate, the water is considered to be impaired. This simple statistical test is strongly biased toward type I error (identifying a water body as impaired when it is not [U.S. EPA, 1997]). On the other hand, the raw-score-test approach tends to minimize type II error for a particular data set, making it extremely conservative—that is, tending toward not "missing" waters that are impaired. The WERF (2003b) found that several surveyed states favored this approach, but that some of them were reconsidering the approach. The deficiencies of this approach to water-quality assessments and the TMDL process are presented by Smith et al. (2001). Gibbons (2003) offers a comparison of several statistical treatments, comparing them to each other and to U.S. EPA's method. He found that the U.S. EPA approach diverged most dramatically from other statistical methods for small sample sizes and for values close to the standard. These conditions are quite prevalent in the real world.

Several alternatives to the raw-score test are available (Gibbons and Coleman 2001; Smith et al., 2001). In general, they analyze the probability of a water-quality standard exceedance, based on the data. They also have the significant attribute of allowing the water-quality manager to decide explicitly the tolerable probability of type I and II errors. This decision process would weigh the risk of type I errors, such as the cost of unnecessary water-quality studies or pollution-control facilities, against the risk of type II errors, such as human contact with disease organisms or damage to aquatic biota.

The WERF (2003b) reports that several states, including Florida, Mississippi, Texas, and Washington, specify a minimum number of samples for assessing water quality. Florida is presented as an example. The state has established a pair of tables (one for planning and one for verification) that define the number of exceedances of a standard that must be observed to consider a water body impaired.

Florida Department of Environmental Protection: Number of Samples Needed to Assess Criterion Exceedance

The WERF (2003b) reprints tables developed by Florida's DEP, based on binomial testing of data to identify a 10% exceedance at the 80% confidence level (that is, one would be 80% confident that an exceedance of 10% of a standard value will be detected) for planning-level decisions regarding listing waters as impaired. The Florida DEP also provides tables that cover the verification step of the listing process, where the confidence level is 90%.

The effect of confidence level on the number of exceedances required to support a listing decision are illustrated in Table 6.1. (Note: Complete lists can be found in WERF, 2003b.)

DATA ISSUES CONTINUED

U.S EPA has responded to concerns regarding the disparate approaches of water-quality regulatory agencies in a document titled *Consolidated Assessment and Listing Methodology* (CALM) (U.S. EPA, 2002), which offers comprehensive guidance on developing and implementing data collection programs to support water-quality decisionmaking. The document, which references the work of ITFM (ITFM, 1995), stresses the value of statistically valid information derived under clear data-quality objectives and includes a clearly written discussion of statistical methods in its Appendix C. It also cites the "10% exceedance criterion", implying that regulators should adopt approaches that are less likely to yield excessive type I errors. Sample size is stressed as among the most important aspects of a valid data-collection effort; a sample size of 30 or more sampling units is recommended. U.S. EPA states that, "(I)n the overwhelming majority of water quality standards scenarios, an approach based on probability sampling, in which states define an acceptable probability of decision error, will be preferred" (U.S. EPA, 2002). However, this preference does not take the form of a conclusive guideline for state regulators, and a general concern that

TABLE 6.1   Exceedances needed to list waters as impaired (WERF, 2003b).

NUMBER OF SAMPLES	PLANNING	VERIFICATION
20	4	5
40	7	7
60	9	10
80	11	13
100	13	15

regulators should not "miss" conditions of impairment (type II errors), while reshaping statistical approaches to reduce Type I errors, seems to pervade the document.

In a guidance document prepared by U.S. EPA following the Sections 303(d) and 305(b) (CWA, 1972) reporting cycle in 2002, the agency continued its effort to provide a framework for the integrated report to be prepared by the states in the 2004 cycle (U.S. EPA, 2003). U.S. EPA encouraged states to adopt and describe methodologies that are "consistent with sound science and statistics". Referring to CALM (U.S. EPA, 2002), the guidance calls for states to clearly describe monitoring and evaluation practices for watershed stakeholders, particularly in cases lacking sufficient data or when data "do not meet optimum conditions." On the other hand, U.S. EPA seems to admonish the states against excluding data from a listing-decision process when such exclusion might cause the state to decide not to list a water body as impaired. A similar thread of concern for "missing" a potential condition of impairment seems to run through the document. However, in the end, U.S. EPA appears to offer Category 3—insufficient data to determine whether any designated uses are met—as a decision alternative for the states, with this category requiring states to "schedule monitoring on a priority basis to obtain data...to move these waters into Categories 1, 2, 4, and 5" (U.S. EPA, 2003). Through this circuitous route, the agency may be concurring in an adaptive approach to water-quality management.

U.S. EPA's 10% guidance probably results in the listing of many water bodies as impaired that might not be listed if water-quality data were tested by means less biased toward type I errors. One possible approach to easing the potential burden of such errors would be to revisit the data and subject it to more rigorous statistical testing. However, one would find that many listing decisions made by the states were not supported by reliable data. In a March 2000 report, the GAO noted that only 19% of the nation's rivers and streams and 6% of ocean and shoreline waters were actually assessed by the states to produce the 1996 National Water Quality Inventory (U.S. EPA, 1996). Further, GAO noted that most water-quality-monitoring programs used by the states to identify impaired waters do not provide sufficient data of a quality that would support valid statistical evaluation (GAO, 2000). In other words, few data sets exist, and those that do exist would not be very useful.

The general lack of satisfactory data should not be a surprise. For most of the 30-plus years since enactment of the CWA (1972), U.S. EPA and the states have concentrated on controlling point-source discharges by promulgating technology-based effluent guidelines for wastewater generators and implementing the NPDES permit program. Most water-quality-sampling programs have focused on point sources; the dischargers sampled and tested their discharges, and the states sampled and tested waters in the vicinity of those discharges. Relatively little sampling and testing was done to define ambient water-quality conditions beyond the immediate vicinity of effluent discharges.

IMPLICATIONS FOR WASTEWATER TREATMENT PLANT MANAGERS

Familiarity with monthly discharge monitoring under NPDES should not lull the WWTP into a feeling of complacency because these data may take on additional importance in a water-quality-trading regime. For example, the Connecticut nitrogen trading program for Long Island Sound (described in Chapters 2 and 5) allows (or requires) point-source dischargers to purchase credits from the program if they have exceeded the loading caps assigned each year. The statistical protocol is defined by the trading regime (in this case, an annual average of total nitrogen measurements). Thus, the annual nitrogen loading as measured by the discharger will translate directly to a financial effect. If the loading is under the cap, the nitrogen trading regime will buy the difference from the discharger; however, if the loading exceeds the cap, the discharger will have to purchase credits to make up the difference (Johnson, 2004; WERF 2000). While the statistical treatment is defined, the WWTP can and should evaluate its sampling and analytical procedures to identify nonrandom errors that might bias the annual average.

A permit limit expressed as an annual average or a 12-month rolling average offers an example by which to evaluate the factors that affect the monthly average. Unless the WWTP is operating under a prescribed monitoring regime, it should carefully evaluate the plant operation and devise a sampling program that will truly represent the quality of the discharge. An effective program will entail thoughtful planning and ongoing oversight. It is difficult to overemphasize the importance of this activity, particularly as discharge limits become more stringent and as the measured loading comes to be translated into either a financial benefit or a cost.

A 12-month rolling average loading, for any pollutant, will be generated by summing the monthly averages for the prior twelve months and dividing by 12. Each monthly average has an equal effect on the rolling average, and any monthly average will affect the rolling average for one year. If the plant-discharge-monitoring program happens to include a particularly high pollutant value, this record will enter the rolling average computation and persist for one year. It is important to detect elevated pollutant concentration events to accurately present the loadings reaching the receiving waters; however, it is also important to identify the duration (and, therefore, the volume) of effluent discharged during the event. If a routine monitoring program involves a daily analysis for the pollutant of interest and if a process upset at the time of routine sampling results in an elevated pollutant concentration, that concentration will be multiplied by the total volume discharged during that day to yield the pollutant loading. If the duration of the process upset was less than a day, the actual pollutant loading in the discharge would be less than the calculated value; however, in the absence of additional concentration data, the calculated loading would enter the monthly average. If routine monitoring entailed analyses less frequent than daily, the situation could be even more severe. An unexpectedly high pollutant concentration would be multiplied by the entire flow volume treated during the interval between samples, and the higher loading would exert a larger effect on the monthly average because there will be fewer loading values in the computation.

Accurate representation of pollutant loadings requires a thorough understanding of the variation of pollutant concentration and plant flow over time. Variations could be caused by daily, weekly, or seasonal wastewater generation patterns and by storm events. Wastewater volumes should be readily available from plant-flow-metering records, but pollutant concentration data may be sparse. Pollutant concentration should be measured with respect to each of the variables that might influence it, and the evaluation period should be long enough to provide a reliable view of relationships. The evaluation program might also identify relationships between easily measured indicator compounds and those compounds that must be reported in discharge monitoring reports. For example, nitrate, which can be measured quickly with a probe, might serve as an indicator for total nitrogen (the reported value). These relationships will allow the WWTP to respond quickly during unexpected excursions, perhaps by taking additional samples to accurately define the duration of the events.

The pollutant evaluation will allow the WWTP to devise a monitoring program that accurately represents pollutant loadings by setting times for routine sampling and defining supplemental sampling procedures for wet-weather events or for process upsets, as these conditions are encountered. A WWTP would be well-advised to obtain concurrence from the state regulators and to offer the monitoring program for review by other stakeholders in the watershed if an appropriate forum has been established.

Implicit in the use of a simple average to express the annual loading from a point-source discharge is the assumption that the data will vary around that average in a random manner—the familiar normal distribution. However, if nonrandom errors enter into the process, their effect on the calculated average cannot be discerned from the customary expressions of variance. Nonrandom errors might result from consistently wrong sampling procedures or from consistent errors in laboratory procedures. This kind of error is also called bias, which is defined by Gibbons and Coleman (2001) as "systematic or persistent error due to distortion of a measurement process, which deprives the result of representativeness". Although bias cannot be discerned by calculating variance of the raw data, it can often be revealed by some relatively simple graphical analyses. Berthouex and Brown (2002) offer some techniques for plotting raw data or residuals of the data (the differences between data points and the calculated mean) against factors that might introduce bias, such as time of day, day of the week, or identity of the chemist. When bias is identified, the sampling and analytical procedures can be modified to eliminate it.

MODELING ISSUES

If developing a set of data to support rational water-quality management is a challenge, selecting, calibrating, and using a model compounds the challenge. Modeling is the "other half" of water-quality management. Models occupy the center of water-quality-management programs by predicting the effect of adding or removing pollutants from a watershed. They are also used to predict the amount of pollutant that would result from various point-source-control technologies or from nonpoint-source best management practices (BMPs).

Models can be of two general types: mechanistic or empirical. The mechanistic model is based on perceptions of how the aquatic system actually functions, while the empirical model is not necessarily strongly connected to perceptions of the natural system. The empirical model is based on correlations that imply cause and effect, and it may fail on close examination of such implied linkages. Empirical models may also produce false correlations if variables are related in ways not recognized by the model designers. Examples of absurd empirical models can be imagined. (The rooster's crowing causes the sun to rise in the morning; we could argue the cause and effect, based on strong correlation, but we know that the rooster does not cause the sun to rise.) Another data set, such as of observations of the time to reach a destination and velocity, are perfectly correlated because they are related. While an empirical model designer would not blunder into such obvious errors, complex natural systems present numerous opportunities to make inappropriate cause-and-effect assumptions, and they may be rife with hidden relationships.

Most, if not all, of the modeling in water-quality-management programs consists of a number of linked mechanistic model units that seek to portray the physical and biochemical responses of the aquatic system to various conditions and inputs. Although we can take some comfort in making the effort to define natural processes by describing them mathematically, we should bear in mind that we are merely trying to match our observations of the natural system behavior with a somewhat arbitrary selection of a mathematical expression from a library that is limited by practicality. In a very real sense, we are devising a series of small empirical models and linking them according to our perception of how they probably interact in nature. Further, while we can increase our confidence in data by collecting and analyzing more samples; the degree of refinement that we can attain in a model will be limited by practical constraints, such as computational power. A model that would predict interactions in the aquatic environment at the molecular level (if such a model could be developed) would be hopelessly unwieldy for real-world use.

Chapra (2003) provides a lucid description of modeling applied to the TMDL process. He makes a number of important points, recognizing the value and the blemishes of modeling and concluding that, applied thoughtfully, they are essential tools for managing water quality in the "TMDL age". If the age of point-source control seems to have been simpler, it is useful to recall that, in the 30 years after enactment of the CWA (1972), regulators and point-source dischargers became comfortable with (or resigned to) the models that were used to predict the effect of such discharges on receiving waters. Point-source-effect modeling seemed to be routine and simple. In reality, it was simple only because the interested parties generally found the process to be tolerable and reasonably predictable. Actually, modeling of point-source effects on receiving waters was fraught with uncertainty, such as the differing effects of pollutants from a WWTP during dry and wet weather. Often, the parties would be willing to "gloss over" such uncertainties, because low stream-flow conditions probably accurately reflected the period of maximum potential effect by the point source. Further, the regulatory emphasis was often focused on incremental increases in control technology rather than on fine-tuning a treatment process to comply with a particular water-quality standard.

In the effluent-based regulatory environment that existed on most water bodies, the next technological step for a point-source discharger could be defined, and the question of sharing the burden of attaining water-quality standards pertained only to neighboring point sources. Under a TMDL, the prospect of sharing the burden among point and nonpoint sources makes earlier accommodations between regulators and point-source dischargers unrealistic. In this water-quality-based regulatory environment, each watershed presents a unique situation, deserving of a unique solution.

Water-quality models depend on data of adequate quality to provide predictions that are reasonably certain to resemble the truth. Unfortunately, with very few exceptions, the modeling element of water-quality-based-management programs lacks sufficient data to fill this critical role. Many modeling efforts differ little from programs developed to support dry-weather, effluent-based regulation (Freedman, 2001). Models are mathematical representations of natural processes. These mathematical expressions include coefficients, which are used to "tune" the model to fit actual observations. This tuning, which is called calibration, is done with data. As a comprehensive water-body model consists of many smaller, linked models, the number of coefficients that need to be defined through the calibration process can be fairly large. Further, data must be available (or must be generated) to calibrate specific model elements. Such data may be difficult or costly to produce, and the modeler may have to refer to the literature.

Modeling for watershed-based, water-quality management has two general objectives

(1) Predicting the response of a water body to pollution or pollutant input (the water-body model) and

(2) Predicting the pollution or pollutant input from point and nonpoint sources (the watershed model).

The water-body model may consist of a number of elements, such as dissolved oxygen, algae growth, sediment transport, and bacterial fate. The water-body model also must deal with transient events in the watershed. Each of these elements involves coefficients that must be calibrated with data derived for the purpose, or they must be estimated from literature sources. Watershed models may include the familiar point-source-discharge simulation and models to predict nutrient and sediment washoff from agricultural land and urban wet-weather runoff pollutants, such as nutrients and bacteria. These models present more calibration demands, requiring either data or literature values for coefficients.

A detailed evaluation of three modeling approaches was performed as part of the TMDL process for the Neuse River estuary in North Carolina (Stow et al., 2003). Two mechanistic models (two- and three-dimensional) and a Bayesian probabilistic model were run side-by-side, in an effort to determine which model should be used for TMDL decisionmaking. A Bayesian model uses an assumed distribution of values for model variables. These distributions may be derived from prior statistics for the variables in the location of interest, or they may be inferred by the modeler, based on distributions observed in other locations. The Neuse River estuary study had a significant amount of data from monitoring programs that had been carried out over the years. Results of the evaluation were apparently startling to the technical team. "Results from the model verification exercise are humbling," the authors reported (Stow et al., 2003). Additional details of the study are described in the Neuse River Estuary Modeling Study section.

The Neuse River estuary modeling experience represents a precautionary tale for all parties to a water-quality-management undertaking. Unless and until modeling efforts have accumulated a great deal of data derived specifically for the needs of the models, output generated by the models should be considered as indicators of the nature of outcomes from regulatory actions and not as firm predictions. In the words of the authors of the Neuse River estuary study, "(T)he utility of models is to provide quantitative guidance rather than a definitive number" (Stow et al., 2003).

Neuse River Estuary Modeling Study

The North Carolina Division of Water Quality (NCDWQ) developed a phased TMDL for the Neuse River estuary to address chlorophyll *a*, which had been increasing steadily since the early 1970s. The process actually started as the Neuse River Nutrient Sensitive Waters Management Study (NCDWQ, 1997), which NCDWQ developed in 1997. This study, which was not a TMDL, yielded the "Neuse Rules", which required nitrogen loads to be reduced by 30%, based on 1991-to-1995 levels. These rules were subsequently approved by U.S. EPA as the first phase of a TMDL, with the requirement that a second-phase TMDL be prepared, based on data derived from monitoring programs conducted between 1996 and 2000. Phase II included a comparison of three modeling approaches

(1) A two-dimensional, mechanistic model known as the "Neuse Estuary eutrophication model";

(2) A three-dimensional, mechanistic model known as "water analysis simulation program", and

(3) A Bayesian probabilistic model, named the "Neuse Estuary Bayesian ecological response network".

All three models were calibrated using pre-2000 data, and model outputs were compared to observed 2000 chlorophyll *a* measurements.

None of the models were deemed to be able to offer satisfactory performance for the purpose of explicit chlorophyll *a* predictions for all sections of the estuary. The two mechanistic models were also hampered by lack of boundary condition data; they could not be used to simulate the 1991-to-1995 base period. This finding pointed out a subtle and important point; regardless of the amount of historical data available—and the Neuse had an extensive body of data—the data might not support later modeling if certain information was not collected. The modeling team stated, "(O)ur results indicated that, even in a well-studied, data-rich system, accurate prediction is difficult" (Stow et al., 2003).

Although the study concluded that none of the models could be depended on to provide specific predictions for a given hypothetical situation (such as a particular nitrogen loading control regime), models are nonetheless essential for all but the simplest TMDLs. The study team suggested that models should be used in a collaborative atmosphere, with ample stakeholder involvement, to provide "quantitative guidance rather than a definitive number" (Stow et al., 2003).

The Neuse River estuary modeling experience represents a strong endorsement of the adaptive approach to water-quality management. The early "Neuse Rules" turned out to be fairly close to the eventual phase II TMDL decision, and the collaborative process achieved substantial consensus among stakeholders.

MODELING ISSUES CONTINUED

U.S. EPA has assembled a suite of modeling tools into its "Better Assessment Science Integrating Point and Nonpoint Sources (BASINS)" product (U.S. EPA, 2001) to support TMDL development. The package includes national water-quality databases, data extraction tools, an instream water-quality model, and two watershed-loading and transport models. U.S. EPA has purposely established an architecture that allows users to include other program elements, such as extensions, to offer a flexible tool for water-quality modeling. This openness seems to further confirm the agency's trend toward collaboration in watershed-based, water-quality-management efforts. Stow et al. (2003) and Maguire (2003) cited the substantial positive effect on the collaborative process that resulted from stakeholder involvement in the modeling aspect of the Neuse River estuary TMDL process. Despite their considerable limitations, models are essential to all but the very simplest water-quality management efforts.

EFFECT OF SCIENTIFIC UNCERTAINTY ON WATER-QUALITY TRADING

Data and modeling uncertainties will affect pollution-credit transactions in essentially the same ways that they affect water-quality-management programs. The models that are set up to implement the management program would logically be used to predict the benefit of pollution reduction in different locations in the watershed. The trading activity will introduce one more element of uncertainty into the mix—valuation of the credit. Point-source trades might be fairly straightforward. A mass unit of pollutant removed here provides a credit there for a related mass unit. Ratios may need to be established to account for different travel times from the trading point sources to the point of interest, but these ratios do not seem to be controversial. Trades between point and nonpoint sources, however, introduce a pair of questions that will need to be addressed

■ How much pollution will a management practice remove?

■ How will the performance of the management practice be confirmed, in the near term and over the longer term?

While it may be possible for parties to agree to some generalizations regarding the positive effects of management practices, they are likely to be areas of significant debate, at least for the near term, until a body of knowledge has been developed through a valid scientific process.

Nonpoint-source BMP effects might be included in a water-quality model in two general ways. The first would involve modeling of the effect of uncontrolled activities that generate nonpoint-source pollutants using a model such as the soil and water assessment tool (SWAT), which is included in U.S. EPA's BASINS suite, or the annualized agricultural nonpoint-source-pollution model (AnnAGNPS). Pollutant reductions resulting from BMPs would be introduced by applying a factor derived from reported research, perhaps the mean of values reported for a particular practice. The model would be run with BMPs (assumed to be in place) to generate pollutant loadings for the controlled condition (Evans, 2004). An alternate approach, which might be applicable in very simple situations, would be to stipulate the pollutant reduction that would be expected by a BMP unit.

Projections of pollutant contribution by nonpoint sources and pollutant reductions that might be achieved through BMPs introduce some significant potential uncertainties to the water-quality-management process. First, models such as SWAT and AnnAGNPS would probably not be verified by field measurements in the watershed of interest. Site-specific calibration would be prohibitively expensive; nonpoint-source-monitoring programs require significant sampling infrastructure and effort. Thus, in most situations, the models will be applied using values for variables that have been derived from intensive field studies on heavily instrumented watersheds, perhaps at some distance from the watershed of interest. For example, AnnAGNPS is reported by Yuan et al. (2003) to have been verified on watersheds in Mississippi and Minnesota. While this approach is likely to be adequate in most situations, stakeholders must bear in mind that site-specific verification will not be feasible.

The second element of uncertainty related to BMP effectiveness is introduced with the factors chosen to represent pollutant reduction that might be expected when the measures are put in place. The mean values from reported research would seem to be appropriate, but the literature offers only limited results. As research advances in this area, it will be possible to select pollutant reduction factors with more confidence.

As watershed models are run and prospective BMPs are tested, a significant amount of conservatism may be introduced to decisions by modelers. It is important for watershed stakeholders to understand this process because another conservative factor—the trading ratio—is likely to be invoked at some point to the process. As the likely buyer of trading credits, the WWTP would be wise to insist that all of the potential sources of conservatism be considered together because conservative modeling assumptions and the trading ratio will increase the price of those credits.

Summary: Do Not Despair

In the face of uncertainty, in all aspects of water-quality management and pollution credit trading programs, WWTP management might be tempted to surrender to the seemingly limitless chaos and accept what the process brings (or see what happens and then resist). Responding in this way would cause the WWTP to miss the best opportunity to affect process. One should keep in mind that all parties with interests in the outcome of a water-quality-management program (regulators, dischargers, environmental advocacy groups, political bodies, and the public) face the same seemingly chaotic situation. Those parties that resolve to work together to meet plausible water-quality standards that support attainable uses will be able to, at least, air their concerns and opinions. Those parties that avoid the process will probably have to accept what the process imposes on them.

THE HUMAN FACTOR

There is little doubt that the body of knowledge and data for most watersheds are not sufficient to support a rigorous, scientific approach to water-quality management. There is also little doubt that we need to move beyond point-source controls achieved through the NPDES program to achieve our clean-water goals. Society is not likely to accept years of status quo while we gather sufficient data of adequate quality to implement a fully formed, water-quality-management program. Thus, the best approach would seem to be to move ahead deliberately, with all stakeholders offered the opportunity to participate in a scientific endeavor to understand our watersheds and to find practical ways to improve the quality of our waters.

Maguire (2003) presents a cogent description of the interactions of stakeholders in the Neuse River TMDL process. Fourteen rules are proposed to govern the process, organized within the following themes:

- ■ Stakeholder interaction with model development,
- ■ Interaction of scientists and stakeholders,
- ■ Integration of stakeholder values with science,
- ■ Interplay of science and stakeholders with regulatory decisionmaking, and
- ■ Adaptive management.

A number of important successes are identified, along with some persistent drawbacks. Maguire observes that the failures of the process to address some of the most important stakeholder concerns probably resulted from "an overly narrow construction of regulatory decisionmaking, where the scientific basis for regulation is limited to biophysical concepts of the river system, rather than the full range of biological, economic, social, and cultural elements, which should inform regulatory decisions" (Maguire, 2003).

Paraphrasing Maguire's conclusion, we must recognize that water-quality management through the TMDL process is, at its foundation, a political process—and an immature one. Stakeholders come to the table with diverse interests and unbalanced power. The collaborative process is certain to result eventually in decisions regarding our water resources that bear the imprint of all stakeholders. The important point is to keep working to improve decisions.

The Connecticut DEP developed and executed a process intended to inform and obtain input from stakeholders as it prepared its Long Island Sound nitrogen trading program. The process is described in the following section (Johnson, 2004).

Stakeholder Involvement in the Development of the Connecticut Long Island Sound Nitrogen Trading Program

Connecticut's actions, related to nitrogen reductions to the Long Island Sound, were initiated to respond to the development of a TMDL that was ultimately approved by U.S. EPA in April 2001. The Connecticut DEP was empowered by legislation passed in July 2001 to prepare a statewide general NPDES permit to cover nitrogen discharges from 79 publicly owned treatment works, as described in Chapter 5. Connecticut DEP prepared a draft of the permit and carried out a series of public meetings in the fall of 2001 to present it to stakeholders. Six meetings were scheduled and convened, and several more meetings were held with specific municipalities to explain the proposed trading process and to obtain input from affected dischargers and other interested parties. A public hearing was convened, in two parts, in Hartford, Connecticut—the first day in October and the second day in November. Nearly 20 written comments were received.

Thoughtful development of the Connecticut Long Island Sound nitrogen trading concept, involving numerous stakeholders, undoubtedly contributed to the program's success. Regulators, dischargers, and environmental interest groups studied trading alternatives extensively for several years before completion of the TMDL, and the adopted program enjoys wide stakeholder support.

Conclusion

As the WWTP takes its place at the stakeholder table, an understanding of the motivation of the other groups at the table will be useful. The following observations may be helpful:

- ■ The regulator may face pressure to reach an endpoint and may resist a more deliberate, stepwise approach, which draws out the process. This stakeholder probably has to attend to dozens of other watersheds. Pressures may come from agency management, who may need to account to U.S. EPA; this pressure tends to create a desire for closure, to boost agency statistics. Pressures may also come from environmental advocacy groups, which may bring legal action if progress is perceived as too slow.

- ■ Other point-source dischargers will watch for equal treatment among the point sources. Although all are "in the same boat", each has to answer to its respective authorities, and each is going to want to report that the group is being treated equally.

- ■ Environmental advocates want to see certainty of outcome. Because point sources offer the most certainty and point-source controls seem to be more predictable in their effect, they will generally focus on controlling these sources. They will also tend to be skeptical of trading between point and nonpoint sources, as the effects of nonpoint-source-pollution-control practices are difficult to monitor. They will probably seek heavy multipliers in the credit trading equation to account for this uncertainty.

▣ Members of the agricultural community will tend to question their contribution to pollution in the watershed. They will also look for funding of pollution control practices. On the other hand, they will probably tend toward an attitude of environmental stewardship, which may make them strong advocates of methodical programs for improvement.

▣ Political bodies will probably be driven by local economic concerns (jobs and commerce). They will probably tend toward favoring a deliberate approach, seeing it as less disruptive to the regional economy.

▣ Members of the general public are likely to be interested in achieving water- quality goals, but strongly influenced by local economic issues.

The water-quality-management process is, in the end, a political process. We develop models to indicate general directions, but we must always remember that the "science" will always be limited, by funding, but also by practicality. The TMDL program and any associated pollution-credit trading activity should be guided by our models, but decisions as to which parties remove how much pollution need to be made at the stakeholder table. The process is not easy, but nothing of worth ever is.

References

Berthouex, M. B.; Brown, L. C. (2002) *Statistics for Environmental Engineers*; CRC Press: Boca Raton, Florida, 489.

Brown, L. C. (2001) Modeling Uncertainty: QUAL2E-UNCAS Case Study. *Proceedings of the TMDL Science Issues Conference*; St.

Louis, Missouri, March 4–7; Water Environment Federation: Alexandria, Virginia, p 46–51.

Chapra, S. C. (2003) Engineering Water Quality Models and TMDLs. *ASCE J. Water Res. Plan. Manage,* **129** (4), 247–256.

Clean Water Act (1972) U.S. Code, Section 1251–1387, Title 33.

Evans, B., The Pennsylvania State University, University Park, Pennsylvania (2004) Personal communication.

Freedman, P. (2001) The CWA's New Clothes. *Water Environ. Technol.,* **13** (6), 28–32.

Gibbons, R. D. (2003) A Statistical Approach for Performing Water Quality Impairment Assessments. *J. Am. Water Res. Assoc.,* **39** (4), 841–849.

Gibbons, R. D.; Coleman, D. E. (2001) *Statistical Methods for Detection and Quantification of Environmental Contamination*; Wiley & Sons: New York, p 384.

General Accounting Office (2000) *Water Quality: Key EPA and State Decisions Limited by Inconsistent and Incomplete Data*; GAO/RCED-00-54. Report to the Chairman, Subcommittee on Water Resources and Environment, Committee on Transportation and Infrastructure, House of Representatives; U.S. General Accounting Office: Washington, D.C.

Hansen, E. (2001) Cheat River Acid Mine Drainage TMDL Case Study: Increasing Stakeholder Confidence in Computer Models. *Proceedings of the TMDL Science Issues Conference*; St. Louis, Missouri, March 4–7; Water Environment Federation: Alexandria, Virginia, p 199–204.

Hauck, L.; Vargas, M. (2001) A TMDL Allocation of Point and Nonpoint Sources: Are all Loadings Equal? *Proceedings of the TMDL Science Issues Conference*; St. Louis, Missouri, March 4–7; Water Environment Federation: Alexandria, Virginia, p 226.

Intergovernmental Task Force on Monitoring Water Quality (1995) *The Strategy for Improving Water-Quality Monitoring in the United States*; Open File Report 95-742; U.S. Geological Survey: Reston, Virginia.

Johnson, G., Connecticut Department of Environmental Protection, Hartford, Connecticut (2004) Personal communication.

Lee, G. F. (2001) Issues in Developing the San Joaquin River, CA. DO TMDL: Balancing Point and Nonpoint Oxygen Demand/Nutrient Control. *Proceedings of the TMDL Science Issues Conference*; St. Louis Missouri, March 4–7; Water Environment Federation: Alexandria, Virginia, p 255–284.

Maguire, L. A. (2003) Interplay of Science and Stakeholder Values in Neuse River Total Maximum Daily Load Process. *ASCE J. Water Res. Plan. Manage.,* **129** (4), 261–270.

National Research Council (2001) *Assessing the TMDL Approach to Water Quality Management*; National Academy Press: Washington, D.C.

North Carolina Division of Water Quality (1997) *Neuse River Nutrient Sensitive Waters Management Study*; North Carolina Division of Water Quality: Raleigh, North Carolina.

Reckhow, K. H. (2003) On the Need for Uncertainty Assessment in TMDL Modeling and Implementation. *ASCE J. Water Res. Plan. Manage.,* **129** (4), 245–246.

Smith, E. P.; Ye, K.; Hughes, C.; Shabman, L. (2001) Statistical Assessment of Violations of Water Quality Standards Under Section 303(d) of the Clean Water Act. *Environ. Sci. Technol.,* **35** (3), 606–612.

Stiles, T. C. (2001) A Simple Method to Define Bacteria TMDLs in Kansas. *Proceedings of the TMDL Science Issues Conference*; St. Louis, Missouri, March 4–7; Water Environment Federation: Alexandria, Virginia, p 375–377.

Stober, P. E.; Henderson, G.; Tierney, D.; Martin, R.; Taylor, S. (2001) Assessment of Nonpoint Source Dominated Watersheds in Missouri. *Proceedings of the TMDL Science Issues Conference*; St. Louis, Missouri, March 4–7; Water Environment Federation: Alexandria, Virginia, p 379–391.

Stow, C. A.; Roessler, C.; Borsuk, M. E.; Bowen, J. D.; Reckhow, K. H. (2003) Comparison of Estuarine Water Quality Models for Total Maximum Daily Load Development in the Neuse River Estuary. *ASCE J. Water Res. Plan. Manage.,* **129** (4), 307–314.

U.S. Environmental Protection Agency (1996) *National Water Quality Inventory*, 305(b) report; U.S. Environmental Protection Agency: Washington, D.C.

U.S. Environmental Protection Agency (1997) National Clarifying Guidance for 1998 State and Territory Clean Water Act Section 303(d) Listing Decisions. Unpublished guidance. U.S. Environmental Protection Agency, Office of Water: Washington, D.C.

U.S. Environmental Protection Agency (2001) Better Assessment Science Integrating Point and Nonpoint Sources (BASINS); EPA-823/B-01-001; U.S. Environmental Protection Agency: Washington, D.C.

U.S. Environmental Protection Agency (2002) *Consolidated Assessment and Listing Methodology (CALM) Toward a Compendium of Best Practices*; U.S. Environmental Protection Agency, Office of Wetlands, Oceans, and Watersheds: Washington, D.C.; http://www.epa.gov/owow/monitoring/calm.html (accessed February 24, 2005).

U.S. Environmental Protection Agency (2003) Guidance for 2004 Assessment, Listing and Reporting Requirements Pursuant to Sections 303(d) and 305(b) of the Clean Water Act; TMDL-01-03, Memorandum. U.S. Environmental Protection Agency, Office of Wetlands, Oceans, and Watersheds: Washington, D.C.; http://www.epa.gov/owow/tmdl/tmdl0103/ (accessed February 24, 2005).

Wang, L.; Kanehl, P. (2003) Influences of Watershed Urbanization and Instream Habitat on Macroinvertebrates in Cold Water Streams. *J. Am. Water Res. Assoc.*, **39** (5), 1181–1196.

Water Environment Research Foundation (2000) *Nitrogen Credit Trading in the Long Island Sound Watershed*; Project 97-IRM-5B; Water Environment Research Foundation: Alexandria, Virginia.

Water Environment Research Foundation (2003a) *Navigating the TMDL Process: Evaluations and Improvements*; Water Environment Research Foundation: Alexandria, Virginia.

Water Environment Research Foundation (2003b) *Navigating the TMDL Process: Listing and Delisting*; Water Environment Research Foundation: Alexandria, Virginia.

Yuan, Y.; Bingner, R. L.; Rebich, R. A. (2003) Evaluation of AnnAGNPS Nitrogen Loading in an Agricultural Watershed. *J. Am. Water Res. Assoc.*, **39** (2), 457–466.

CHAPTER SEVEN

Societal Requirements for Water-Quality Trading

Introduction

This book generally attempts to guide potential water-quality traders in the wastewater community through the myriad considerations required to develop trades or trading programs. In doing so, it touches on a large number of institutional, administrative, and regulatory issues that the wastewater treatment plant (WWTP) must understand and account for in its trading activities. A studious reader could extract all of these issues and requirements and synthesize a set of principles that represent a larger societal perspective of the proper criteria for judging water-quality-trading programs. The purpose of this chapter is to save the interested reader that trouble by summarizing water-quality trading from this larger perspective. It is also the author's intention that this chapter will be of use to members of the regulatory community or the public, who are approaching trading from this perspective.

The chapter begins by describing criteria that have been developed to assess the desirability and usefulness of water-quality-trading programs and the general program elements that must be present, in some manner, before a trading program would meet the criteria and win regulatory acceptance. It does not attempt detailed discussions of the criteria or how to structure the program elements; it merely summarizes and categorizes them to illustrate their necessity. Readers interested in more detailed treatment of these program elements will find the references listed at the end of the chapter very helpful.

Trading Program Criteria

In 1993, the Minnesota Pollution Control Agency (MPCA) began an assessment of point–nonpoint-source trading and issued a report on its findings in 1997. In its assessment, MPCA developed four fundamental criteria by which to judge the validity and desirability of a proposed point–nonpoint-source trade (CPLS, 1999; Senjem, 1997, cited in Fang and Easter 2003). The criteria are as follows:

- ■ Efficiency. There should be economic benefits from the trade.
- ■ Equivalence. There must be interchangeability between the loads being traded so that the trade would produce

equivalent environmental results (or better) to the no-trade situation. This criterion covers a broad range of water-quality-management and trading issues, including scientific uncertainty, data needs, ambient monitoring, interpretation of water-quality standards, load allocations, and assessment of trade effects.

■ Additionality. The credits sold by a nonpoint source to a point source must be the result of load reductions that would not have occurred in the absence of the trade, that is, they must be in addition to expected no-trade load reductions. This is one of the challenges for point-source–nonpoint-source trading programs: how to avoid detracting from ongoing efforts to control nonpoint-source pollution or, going further, how to use trading to augment them.

■ Accountability. There must be means to ensure that trading programs satisfy the equivalence and additionality criteria and meet all other program requirements. This criterion covers all of the aspects of monitoring and tracking trading programs, insuring compliance, taking enforcement actions where necessary, assessing results, and insuring public transparency and opportunity for input.

One of the major challenges of water-quality trading is that it may be difficult to simultaneously satisfy all four of the criteria (Fang and Easter, 2003). It is clear that the last three criteria (equivalence, additionality, and accountability) can and will often conflict with the first one (efficiency). Hence, the challenge to both traders and regulatory agencies alike will be to find ways to maintain favorable economics in the face of unavoidable program requirements.

Necessary Program Elements

The program elements that have been identified by the U.S. Environmental Protection Agency (U.S. EPA), state regulatory agencies, and others as necessary for trading programs are inventoried here. They are grouped in five broad categories—four corresponding to the four criteria described above and a fifth general category. More detailed discussions of most of these elements can be found in the appropriate chapters throughout the book.

GENERAL PROGRAM ELEMENTS

These program elements form the foundation of any trading program and apply across the four criteria.

Policy Direction

Setting overall trading policy is a function shared by U.S. EPA and the states. U.S. EPA has produced the Water Quality Trading Policy (U.S. EPA, 2003), and states are using it to help them design trading programs. The primary requirement in the Water Quality Trading Policy is that trading programs must be consistent with and integrated to the nation's core water-quality-management programs and be fully consistent with all requirements of the Clean Water Act (CWA, 1972). States must integrate trading into state water-quality-management regulations and programs in a way that is consistent with this requirement.

Program Design

Within the general requirements set forth by the trading policy, states have a great deal of flexibility in the design of their trading programs. States will assess the feasibility and desirability of trading, decide whether or not they want to have trading, define the pollutants that are tradable, and establish eligibility requirements for parties wishing to trade. States will design trading programs and produce trading rules in some manner, whether through guidance, regulation, or legislation.

There will be many challenges involved in the assessment of trading and design of trading programs. Two challenges that are not immediately apparent are externalities and moral hazards. They were identified as trading-program-design issues by the Center for Public Leadership Studies (CPLS) in a policy review on water-quality trading prepared for the Texas Natural Resource Conservation Commission (CPLS, 1999). An externality is an effect that a trade would have on a person or entity not party to the trade, that is, an unintended consequence. The creation of a local water-quality hotspot by a trade would be an example of a negative externality. Unintended consequences are, by nature, often difficult to predict. Trading programs and proposals should be carefully assessed for any potential unintended consequences.

A moral hazard is the creation of incentives for traders to engage in behaviors harmful to society as a whole, such as "gaming" the system to maximize profits, while avoiding producing benefits. An example cited by CPLS would be a farmer who increased the nutrient runoff from his farming operations, knowing that doing so would produce more credits for him to sell in the future when he reverted back to his previous practice (CPLS, 1999; p 26). The same thing could potentially be done by any type of discharger, including WWTPs. As with externalities, care must be taken with the trading-program design to avoid creating moral hazards.

Public Input and Transparency

For trading programs to succeed and be accepted by the public as a worthwhile tool for water-quality management, they must be developed with complete public transparency, public education, and adequate opportunities for public comment. The Chesapeake Bay Program Nutrient Trading Fundamental Principles and Guidelines established, as a fundamental principle, that "the involvement of a diverse group of stakeholders must be sought in the design and implementation of state trading programs and related public education initiatives" (U.S. EPA, 2001). An analysis of air, water, and wetlands trading programs prepared for the National Academy of Public Administrators presented a broad list of concerns from diverse groups of stakeholders involved in all three types of trading programs (Kerr et al., 2000; p 71). Typical concerns include the following:

- Is there equitable treatment of all participants?
- Is there recognition of the difference between those that have made much and those who have made little environmental progress in the past?
- Do trading rules produce environmental equivalence or progress?
- Are there clear enforcement procedures?
- Is there sufficient reliable information to evaluate the results of the trading program?
- Does the trading program adequately guard against localized or temporal environmental effects?

Trading programs and the rules developed to govern them must address these public concerns and any others that the public or participants identify.

Program Evaluation and Revision

In its list of "Common Elements of Credible Trading Programs", U.S. EPA's Water Quality Trading Policy included program evaluations (U.S. EPA, 2003). The trading policy states that "periodic assessments of environmental and economic effectiveness should be conducted and program revisions made as needed" (U.S. EPA, 2003).

The program evaluations will essentially occur on two levels. On one level would be the ongoing programmatic activities needed to assess whether the criteria of equivalence, additionality, and accountability are being met. On a higher level would be the overall assessment of the efficacy, usefulness, and desirability of the trading program.

Water-quality trading is evolving; hence, there is little doubt that program evaluations will be of great interest and revisions will be frequently needed.

PROGRAM ELEMENTS TO ENSURE EQUIVALENCE

The following sections describe the program elements that deal with the technical aspects of water-quality-trading programs and the science of water-quality management.

Data Sufficiency

Adequate amounts of reliable water-quality and pollutant loading data are needed to determine existing water-quality conditions and predict changes that would occur under different management and trading scenarios. Following the implementation of a management measure or a trade, additional ambient water-quality data may need to be collected to assess its effect. U.S. EPA's *Draft Framework for Watershed-Based Trading* identified, as one of its principles of trading, that "trading will generally add to existing ambient monitoring" (U.S. EPA, 1996; p 2–8).

Analytical Tools

Once sufficient data are assured, then acceptable analytical tools to assess existing water-quality and the potential effects of trades must be selected or developed. These tools must be the same as, or consistent with, the tools used to develop total maximum daily loads (TMDLs) or National Pollutant Discharge Elimination System (NPDES) permit limits. This introduces a wide range of technical issues regarding which tools are best and the reliability of the results. Water-quality-model selection, calibration, verification, and model reliability are good examples of these issues. However, it should also be noted that water-quality managers constantly deal with these same issues, with or without trading.

Assessing Water-Quality Results

The monitoring and analytical strategies must also assess the water-quality results of trades at all locations where water quality may be affected by the trade. This includes checking for negative externalities, such as the development of water-quality hotspots in the receiving water of the user of credits.

Dealing with Uncertainty

As noted in Chapter 3 and discussed, in detail, in Chapter 6, scientific uncertainty will always be present in water-quality management. Water-quality practitioners continually strive to improve the science of water-quality management and account for or reduce the uncertainty. Uncertainty must be addressed in developing water-quality-trading programs; however, in general, these programs should be subjected to the same standards for certainty as other water-quality-management programs.

It is interesting to note that, when water-quality trading is proposed, there are frequently calls for it to be subject to a much higher standard of scientific certainty than is applied to virtually any other water-quality-management program. Frequently, there are little objective grounds for the demands; however, there are times when trading programs raise issues of certainty that are not seen in other water-quality-management programs. The most important of these is probably the trading of point-source-pollutant-load reductions, quantified to a high degree of certainty, for poorly quantified and uncertain nonpoint-source-load reductions. This issue (and ways to deal with it) is discussed in Chapter 5.

PROGRAM ELEMENTS TO ENSURE ADDITIONALITY

It is assumed that this criterion was meant by the MPCA to apply to point–nonpoint-source trading. It is logical and necessary to be concerned about it in that context; otherwise, trading could erode existing nonpoint-source-control programs and convert those pollutant reductions to increases in point-source loads. In point-source–point-source trading programs, however, there is no danger of any such thing happening; both trading partners must meet allocations set to achieve water-quality standards, and trading could not result in an increase in loads. Hence, there would be little justification for requiring additionality.

Program elements to ensure additionality in point–nonpoint-source-trading programs are more difficult to generalize about or to design than other elements. Additionality could be endangered in many different ways; some of the ways are probably difficult to foresee. This is made more difficult by the fact that there is inadequate regulatory control over nonpoint sources to ensure that additionality is not endangered. Finding ways to craft trading programs so that nonpoint-source reductions sold as credits to other dischargers are in addition to those needed to meet a water-quality standards or TMDL load allocation is one of the great challenges of trading. It is embodied in the sensitive and tricky issue of establishing minimum requirements or baselines for nonpoint sources to meet before they could participate in trading. This issue is discussed in Chapter 5.

PROGRAM ELEMENTS TO ENSURE EFFICIENCY

It would seem that this criterion would be of greater interest to the regulated community than to regulators, who by virtue of the law and their mission will put more emphasis on issues such as additionality and environmental equivalence. While regulatory agencies are interested in encouraging and supporting efficiency by WWTPs in meeting their permit requirements, they are also interested in the efficiency of their own administration of water-quality regulation. Because a water-quality-trading program could reduce the administrative burden on the regulatory agency, trading programs will be judged by both types of efficiencies.

Efficiency should not be an absolute requirement, however. If two WWTPs wanted to trade because of construction timing, convenience, or some other non-economic reasons, it seems that the trading program should not preclude the trade solely on the grounds that the overall cost would be the same with or without the trade. It must also be said that lower costs are one of the reasons for the interest in water-quality trading, and trading programs should be developed in ways that facilitate this goal.

PROGRAM ELEMENTS TO ENSURE ACCOUNTABILITY

The program elements to ensure accountability can be broken down into two categories: determination of compliance and enforcement.

Determination of Compliance

When dealing with trades between WWTPs, determining compliance would be a simple and straightforward matter. Existing CWA and NPDES mechanisms would be used, notably the self-monitoring and reporting requirements placed on WWTPs that invariably produce voluminous quantities of reliable data. Tracking and verification of the generation and use of credits by WWTPs would be done through the monthly discharge monitoring reports that the plants are required to submit.

This will not be so easily accomplished with point–nonpoint-source trades, however. Ways must be developed to verify the generation of credits by nonpoint sources and track their use, including procedures for credit certification and periodic on-site inspections to verify that best management practices are being properly maintained and agreed-upon practices are being followed. This will not be as easy as with point sources largely because the CWA (1972) does not give regulators any authority over nonpoint sources. State water-quality-trading programs could address this problem, as Michigan has done in its trading rules: any source that generates a load reduction and files a notification with the state to register and certify the tradable credits automatically becomes subject to the enforcement provisions of the regulation (Michigan Department of Environmental Quality, 2002).

Enforcement

Trading programs must include provisions for taking enforcement actions against trading parties that do not meet their obligations. For point sources, enforcement mechanisms should be consistent with the NPDES regulations, as stated in the Water Quality Trading Policy (U.S. EPA, 2003). Enforcement mechanisms for nonpoint sources could be handled in a variety of ways. One good approach is that taken by the Michigan rules in which the regulations confer enforcement authority over anyone participating in trading and also stipulate remedies and penalties in the event of noncompliance. Hence, trading participants must essentially agree to be subject to the legal liabilities established by the regulations or they cannot participate in trading.

References

CPLS (1999) Effluent Trading: A Policy Review for Texas. Report prepared for the Texas Natural Resource Conservation Commission by the Center for Public Leadership Studies: Texas A&M University, College Station, Texas.

Clean Water Act (1972) U.S. Code, Section 1251–1387, Title 33.

Fang, F.; Easter, K. W. (2003) Pollution Trading to Offset New Pollutant Loadings—A Case Study in the Minnesota River Basin. Presented at the American Agricultural Economics Association Annual Meeting, Montreal, Canada, July, 2003; http://www.envtn.org/docs/MN_case_Fang.pdf (accessed March 19, 2004).

Kerr, R. L.; Anderson, S. J.; Jacksch, J. (2000) Crosscutting Analysis of Trading Programs—Case Studies in Air, Water and Wetlands Mitigation Trading Systems. Research paper prepared for the National Academy of Public Administration: Washington, D.C. http://www.napawash.org/pc_economy_environment/epafile06.pdf (accessed April 10, 2004).

Michigan Department of Environmental Quality (2002) Rule Part 30: Water Quality Trading (effective November 22, 2002).

Michigan Department of Environmental Quality, Surface Water Quality Division: Lansing, Michigan; http://www.state.mi.us/ orr/emi/arcrules.asp?type=Numeric&id=1999&subId=1999%2 D036+EQ&subCat=Admincode (accessed April 16, 2004).

Senjem, N. (1997) Pollution Trading for Water Quality Improvement—A Policy Evaluation. Unpublished report, Minnesota Pollution Control Agency: St. Paul, Minnesota.

U.S. Environmental Protection Agency (1996) *Draft Framework for Watershed-Based Trading*; EPA-800/R-96-001; U.S. Environmental Protection Agency, Office of Water: Washington, D.C.; http://www.epa.gov/owow/watershed/trading/framwork.html (accessed June 20, 2004).

U.S. Environmental Protection Agency (2001) *Chesapeake Bay Program Nutrient Trading Fundamental Principles and Guidelines*; EPA-903/B-01-001; Chesapeake Bay Program, Nutrient Trading Negotiation Team: Rockville, Maryland; http://www.chesapeakebay.net/trading.htm (accessed June 20, 2004).

U.S. Environmental Protection Agency (2003) Water Quality Trading Policy. Unpublished guidance; http://www.epa.gov/owow/ watershed/trading/finalpolicy2003.html (accessed June 20, 2004).

CHAPTER EIGHT

Gaining Public Acceptance

Introduction

Public transparency is a key component of water-quality management in the United States. Public review and input are sought at virtually every step of the way, from promulgating federal and state regulations, to developing total maximum daily loads (TMDLs) or watershed plans, to issuing National Pollutant Discharge Elimination System permits. It is this openness that allows all interested parties, including the public and wastewater treatment plants, to influence water-quality-management policies and actions.

Water-quality trading is a relatively new tool for water-quality management. As such, it is not well understood, particularly by the public. Frequently, there are misconceptions about what it entails, often resulting in significant barriers to establishing trading programs. More than with established water-quality-management activities, trading program developers must work hard to overcome misconceptions and skepticism to gain public support. Based on experiences with trading projects across the country, partnerships and alliances among diverse stakeholders are often necessary to establish and ensure continuing support for water-quality-trading programs. This necessity was formalized in the *Chesapeake Bay Program Nutrient Trading Fundamental Principles and Guidelines* as one of eight fundamental principles of trading program design: "The involvement of a diverse group of stakeholders must be sought in the design and implementation of state trading programs and related public education initiatives" (U.S. EPA, 2001).

In its Water Quality Trading Policy, the U.S. Environmental Protection Agency (U.S. EPA) stated that it "...supports public participation at the earliest stages and throughout the development of water-quality-trading programs to strengthen program effectiveness and credibility" (U.S. EPA, 2003). It is unlikely that U.S. EPA would support any trading program that lacked such public involvement.

This chapter addresses critical components for successful public participation in trading-program development. It first addresses the objections and barriers to water-quality trading that are typically encountered. It then discusses which stakeholders should be involved in developing a trading program and the types of attitudes that may be encountered among them. A set of "ground rules" for conducting effective public participation efforts is presented, followed by a set of guidelines based on lessons learned from the Kalamazoo River Basin Phosphorus Trading Program (Kieser, 2000) and other trading projects.

Typical Objections and Barriers to Water-Quality Trading

Typical public objections to water-quality trading may include the following:

- Trading is a way for dischargers to evade their pollution-control responsibilities;

- Trading participants may "game" trading systems, resulting in increases in pollutant loads;

- The circumstances under which trading may be allowed must be greatly restricted (e.g., only after a TMDL has been implemented and its goals achieved); and

- Trades will result in the creation of local hotspots of pollutants.

Typical barriers to public acceptance of water-quality trading may include the following:

- Conflicting perceptions of trading may be held by various stakeholders, including industry, environmental groups, municipalities, farmers, institutions, industry, and the public at large;

- A lack of understanding about who pays and who benefits from trading;

- A general lack of specific water-quality information about the water body where trading is envisioned; and

■ A lack of consensus over water-quality planning efforts and/or a lack of intergovernmental communication and cooperation. This can be even more difficult when watershed boundaries do not coincide with jurisdictional boundaries.

While these are typical objections and barriers, there may be others unique to the locality or watershed. Program developers should ascertain these objections and barriers early in the process.

Identifying and Involving Stakeholders

To overcome misconceptions and objections that may arise, input from a broad array of stakeholders should be sought, from the earliest stages of trading program development. Critical stakeholders may include the following:

■ Watershed groups;

■ Environmental organizations;

■ Conservancy organizations;

■ Citizens and citizens' groups;

■ State regulatory and natural resource agencies;

■ U.S. EPA and other federal agencies;

■ Local or county governments;

■ Farmers;

■ Farm bureaus;

■ Soil and conservation districts;

■ Industry;

■ Wastewater treatment plants; and

■ Consultants.

Outreach efforts should be made to all relevant stakeholder groups. They should be invited to participate in the development of the trading program at the appropriate level, whether it involves periodically providing input on proposed elements, reviewing and commenting on all aspects, or assisting with program development.

Once outreach efforts are completed and a stakeholders group "convened" in some manner, there is likely to be a variety of attitudes toward trading within the group. There may be advocates for trading who want to push rapidly ahead or opponents who hold some of the negative misconceptions noted above. Some stakeholders may not know enough about trading to initially lean one way or another but are there to learn. Past experience has taught that there is likely to be a fair amount of skepticism in the early stages—skepticism that can be overcome only by education and open discussion of the issues.

Given this diversity of opinion (some of it potentially very strong), it is important that the program organizers ensure that the discussions proceed in a harmonious and productive way. It may be necessary to establish ground rules for meetings, such as those established for the Kalamazoo River trading demonstration project, as described in the following section (Kieser, 2000).

Stakeholder Involvement Ground Rules Adopted in the Kalamazoo River Trading Demonstration Project

- Anyone could express him or herself openly in professional discourse;

- Any thoughts or comments relevant to the concept of trading would be listened to and an open, receptive atmosphere would prevail;

- All participants were recognized as "environmentalists" or "friends of the environment" either by advocacy, occupation, education, or any combination of the foregoing;

- Emphasis was placed on members participating as equals;

- Opportunities for productive relationships and partnerships were continuously explored; and

■ Potential agricultural partner sites desiring owner anonymity would receive it rather than risk being identified by name during program development deliberations.

Guidelines for Successful Stakeholder Involvement

Based on the lessons learned from the Kalamazoo River program and other trading projects across the country, a set of guidelines for successful stakeholder involvement can be identified.

LEADERSHIP

Whether stakeholder involvement is handled by an ad hoc committee or a formal organization, good leadership or facilitation is critical for such things as handling the diversity of personalities and backgrounds likely to exist in the group, enforcing the ground rules, and ensuring that the group's activities are well-organized and productive. Ideally, leaders or facilitators should be selected who would be viewed by the group as open, inclusive, honest, reliable, and committed to following through on promises. In addition, a good support system is needed for the leader and the group, as a whole. Without such assistance, the leader is unlikely to be able to perform all that is required over the course of program development.

COMMUNICATION

Because water-quality trading will be a new concept to many stakeholders, it will be important to explain it in a manner that would allow diverse stakeholder groups to not only understand it, but to appreciate the decisions that must be made in the design of the trading program and enable them to participate in making these decisions. Clear and concise explanations should be developed, and communication efforts should be carefully conceived for each audience. With new and, often, unfamiliar terminology being used, it is vital to define terms and concepts as they are first presented and ensure that the audience understands them before moving on.

STAKEHOLDER INTERESTS AND MOTIVES

It is important to understand the interests and motives of all stakeholder groups. Once this insight is gained for a given group, the information provided to them can be tailored to their needs, and communications in general with the group are likely to improve. All points of view should be welcome in the process. An openness to all input, whether supportive, skeptical, or opposed, is absolutely necessary. Without it, stakeholders cannot develop any trust in the process, and, without trust, the skeptics or opponents cannot be won over.

COMMON GOALS

Even with a large and diverse group of stakeholders, there are likely to be many common goals among the members. A good way to begin the process of achieving consensus on a water-quality-trading program would be to start by identifying goals and objectives shared by all participants. These are generally overarching goals, such as improving water quality, increasing recreational opportunities, preserving habitat, and reducing costs. These types of goals are rarely controversial. The agreed-upon goals can then serve as the starting point for the more difficult challenge of determining the best ways to achieve them. The exercise of identifying these common goals can also increase the level of trust, as stakeholders and organizers come to understand how much they have in common.

CONSENSUS AND PROGRESS

It is not likely that full agreement will be reached on all aspects of the trading program or even of the water-quality-management program of which it is a part. Knowing that full agreement is unlikely, the disagreements should not be allowed to stymie the process. The leaders or facilitators should help the participants understand that consensus is the goal and that compromise may be necessary at times, or, perhaps more importantly, participants may, on occasion, simply have to "agree to disagree" on a given issue. Keeping the common goals in mind, the group should remain committed to progress in the face of such disagreements.

DOCUMENT DECISIONS

To maintain full public transparency, it is important that decisions that are made and actions that are taken throughout the development of the trading program be clearly documented, and that the documentation be disseminated to appropriate stakeholder groups and the public. The documentation can be in the form of technical reports, committee and work group minutes, or even public-service announcements. Publicly disseminating this information also provides the opportunity for dissenting groups that may have chosen not to participate to openly track the progress of the project. This transparency would make the project less subject to misrepresentation or unfounded accusations.

Conclusion

Transparency and honest dealings are the cornerstones for establishing trust, and trust is the key to a successful stakeholder involvement effort. All of the different aspects of gaining public acceptance for a trading program that are discussed above are contingent on the building of trust between all the parties involved. When there is mutual respect between participants, and when trust cements the partnerships, the project is bound to succeed. Every trading program will be a little different, depending on the local circumstances, but what should remain constant is that every stakeholder involved feels that her or his voice has been fairly heard and respected (even if their point of view does not prevail). Equally important is that once the trading program is implemented, the participation and contributions of all those involved are acknowledged.

References

Kieser, M. S. (2000) *Phosphorus Credit Trading in the Kalamazoo River Basin: Forging Nontraditional Partnerships;* Water Environment Research Foundation: Alexandria, Virginia.

U.S. Environmental Protection Agency (2001) *Chesapeake Bay Program Nutrient Trading Fundamental Principles and Guidelines;* EPA-903/B-01-001; Chesapeake Bay Program, Nutrient Trading Negotiation Team: Rockville, Maryland, http://www.chesapeakebay.net/trading.htm (accessed June 20, 2004).

U.S. Environmental Protection Agency (2003) Water Quality Trading Policy. Unpublished guidance; http://www.epa.gov/ owow/watershed/trading/finalpolicy2003.html (accessed June 20, 2004).

CHAPTER NINE

Making the Decision

Introduction

Water-quality trading is a new endeavor for many in the wastewater community. This book seeks to facilitate the understanding of how trading programs can fit into and augment the existing water-quality-management programs that govern the activities of wastewater treatment plants (WWTPs). It also attempts to provide perspectives and guidance that go beyond the mere recitation of legal and regulatory policies and requirements. It does so to help WWTPs develop insights into the strengths and weaknesses of the core water-quality-management programs, to identify the opportunities, perhaps even the obligations, to influence these programs in positive ways, and to stress the openness to innovation in water-quality management that is evident throughout the United States today.

Water-quality trading presents, at a minimum, an opportunity for a WWTP to meet regulatory requirements at less cost. Beyond that, there are a host of potential environmental and societal benefits. Ultimately, each WWTP will evaluate this opportunity itself and decide if it wants to participate in water-quality trading.

The Trading Checklist

This book identifies the issues involved in making the decision of whether to trade and the various analyses the WWTP should undertake before deciding. It identifies pitfalls and ways for WWTPs to solve trading problems and minimize risks. It presents information on existing trading programs and options for trading program design.

When a WWTP has finished its assessment of water-quality trading, it should be able to answer the following questions:

- ■ What is water-quality trading?
- ■ What types of water-quality-trading programs have been implemented and/or are available for consideration in the watershed? Across the United States?

- What are the core requirements of the nation's water-quality-management programs into which trading programs must be integrated?

- What are the water-quality goals in the watershed that trading might help achieve?

- What are the costs and cost-related factors that need to be considered in evaluating trading?

- What are the WWTP's trading needs, in terms of quantity and duration?

- How can the trading options be evaluated and a decision made of whether to trade?

- What types of trading partners are needed, and are there sufficient numbers of them in the watershed?

- What special problems must be considered in trading with nonpoint sources? What are some approaches for dealing with them?

- Would credits exchanges be done on a one-to-one basis, or would trading ratios be required? If required, what effect on the costs and benefits of the trading program would they have?

- Would there be multiple benefits resulting from the trade, and could credits be awarded for them?

- Would there be any potential for trades to produce local pollutant hotspots? If so, how could the program be designed to avoid them?

- What type of trading contracts or agreements should be used? What should they contain?

- How would trades be incorporated to National Pollutant Discharge Elimination System permits? What are the risks involved in doing so, and how can they be minimized? What innovative permitting strategies might be used?

- What data and analytical tools are needed to assess proposed trades?

- ■ How can the scientific uncertainty inherent in water-quality management be dealt with in the trading context?

- ■ How well would a trading program satisfy the criteria of efficiency, equivalence, additionality, and accountability?

- ■ What public education and stakeholder involvement efforts would be needed to gain acceptance of the trading proposal? Which groups and stakeholders need to be involved?

While these questions may not be the complete list of those that must be answered, they should be sufficient for the WWTP to make the decision of whether to proceed with trading or to propose a trading program to the state. The WWTP should also carefully consider its unique situation and determine if there are other questions that would be critical for it to answer in making the decision. For additional help, a list of water-quality trading resources is provided in Appendix E.

Conclusion

The authors of this book truly believe that we are experiencing a new era of flexibility and innovation in water-quality management. We have seen the dedication and vigor with which U.S. EPA, state regulatory agencies, and others across the country, such as the Water Environment Research Foundation (Alexandria, Virginia), have worked to develop and promote these innovations. We close by recalling the statement by G. Tracy Mehan that Dave Batchelor cited in the Foreword.

Water-quality trading is an idea whose time has come.

Tracy Mehan is right, and it is now up to us. We strongly urge the wastewater community to rise to the challenge and help further the development of water-quality trading. The wastewater community and society have much to gain.

ACRONYMS

AMSA	Association of Metropolitan Sewerage Authorities
AnnAGNPS	Annualized agricultural nonpoint-source-pollution model
ANPRM	Advanced Notice of Public Rule Making
APAP	Agricultural Pollution Abatement Plan
ASCE	American Society of Civil Engineers
AU	Hydrologic assessment unit
BASINS	Better Assessment Science Integrating Point and Nonpoint Sources
BMP	Best management practice
BNR	Biological nutrient removal
BOD	Biochemical oxygen demand
BOD$_5$	Five-day biochemical oxygen demand
BSL	Base-soil loss
CAA	Clean Air Act
CAFO	Concentrated animal feeding operation
CALM	Consolidated Assessment and Listing Methodology
CBOD$_5$	Five-day carbonaceous biochemical oxygen demand
CGV	Congress Group Ventures
CPLS	Center for Public Leadership Studies
CRWP	Connecticut River Watch Program
CWA	Clean Water Act
CWF	Clean Water Fund
CWS	Clean Water Services
CZARA	Coastal Zone Act Reauthorization Amendments of 1990
DEP	Department of Environmental Protection
DEQ	Department of Environmental Quality

ETN	Environmental Trading Network
FSA	Farm Service Agency
GAAMPs	General accepted agricultural and management practices
GAO	General Accounting Office
ITFM	Intergovernmental Task Force on Monitoring Water Quality
KY DOW	Kentucky Division of Water
MOS	Margin of safety
MPCA	Minnesota Pollution Control Agency
MS4	Metropolitan separate storm sewer
MSD	Metropolitan sewer district
MUA	Multiattribute utility analysis
NAPA	National Academy of Public Administration
NASCD	National Association of Soil Conservation Districts
NCAB	Nitrogen Credit Advisory Board
NCDWQ	North Carolina Division of Water Quality
NPDES	National Pollutant Discharge Elimination System
NRC	National Research Council
NRCS	Natural Resource Conservation Service
NWQMC	National Water Quality Monitoring Council
O&M	Operations and maintenance
POTW	Publicly owned treatment work
PRF	Pollution reduction facilities
SCC	Soil Conservation Commission
SISL	Surface-irrigation soil loss
SO$_2$	Sulfur dioxide
SRF	State revolving fund

STORET	STOrage and RETrieval (U.S. EPA water-quality database)
SWAT	Soil and water assessment tool
SWMP	Stormwater management program
TMDL	Total maximum daily load
TSS	Total suspended solids
U.S. EPA	U.S. Environmental Protection Agency
USDA	U.S. Department of Agriculture
WEF	Water Environment Federation®
WERF	Water Environment Research Foundation
WQBEL	Water-quality-based effluent limit
WQMP	Water-quality-management plan
WWTP	Wastewater treatment plant

GLOSSARY

303(d) list—The list of impaired waters prepared by the states, as described in Section 303(d) of the Clean Water Act.

305(b) report—A report defined by Section 305(b) of the Clean Water Act that the U.S. Environmental Protection Agency must prepare and submit to Congress, based on information provided by the states' 303(d) lists.

Antidegradation—Provision of the Clean Water Act that prohibits actions that would degrade a water body with respect to its current condition.

Antibacksliding—Provision of the Clean Water Act that prohibits less stringent pollutant discharge limits in an National Pollutant Discharge Elimination System permit with respect to an earlier permit.

Assimilative capacity—The capacity of a water body to absorb a pollutant without degrading the water with respect to its designated use.

Banking—The setting aside of a credit, by the discharger that generated it, for future use in a time period beyond that specified by the analytical framework used to set the discharger's allocation.

Baseline—The total mass-load-per-unit-time of a pollutant that a regulated source may discharge under a permit limit, total maximum daily load wasteload allocation, or other regulatory requirement. The discharger must reduce its discharge of the tradable pollutant below the baseline to generate useable credits.

Best management practice (BMP)—Practices pertaining to nonpoint sources that reduce the pollutant load of surface runoff and subsurface flows from agricultural practices and from urban or developed areas.

Biological nutrient removal (BNR)—Biological wastewater treatment technologies that remove phosphorus and nitrogen through biochemical processes with minimal application of chemicals.

Chlorophyll *a*—A surrogate used to indicate the concentration of algae in a water body.

Clean Water Act (CWA)—The 1972 Federal Water Pollution Control Act and federal legislation to reauthorize the act.

Credit—A unit of pollutant discharge expressed in mass-per-unit-time that is created when a discharger reduces its discharge of the pollutant below its baseline requirement. Once created, credits may be sold or exchanged between the dischargers or sources of the pollutant in the watershed, if a trading program exists.

Discharge allowance—The total mass-load-per-unit-time of a pollutant that a regulated source may discharge under a permit limit, total maximum daily load wasteload allocation, or other regulatory requirement.

Load allocation—The pollutant load allocated to nonpoint-source discharges in the total maximum daily load process.

National Pollutant Discharge Elimination System (NPDES)—The point-source-discharge permitting program defined by the Clean Water Act.

Nonpoint source—A pollutant discharge that cannot be identified as a specific point, such as runoff from agricultural areas.

Offset—An offset is a requirement that a discharger take some offsetting action elsewhere in return for increasing its discharged load or for not decreasing it to comply with a new wasteload allocation.

Point source—A pollutant discharge that can be identified as a specific point (an outfall pipe, for example) that requires a permit under the National Pollutant Discharge Elimination System.

Publicly owned treatment works—The Clean Water Act and regulatory term for publicly owned municipal wastewater treatment plants.

State revolving fund (SRF)—Agency in each state, generally associated with the environmental regulatory agency, that receives and administers federal funding for construction of wastewater treatment systems.

Technology-based effluent limits—Limits established by the U.S. Environmental Protection Agency for pollutant discharges under the National Pollutant Discharge Elimination System program, based on the kind of enterprise generating the pollutant.

Total maximum daily load (TMDL)—Regulatory process described in the federal Clean Water Act that is to be carried out if waters are deemed to be impaired, in spite of application of effluent limits on all point-source dischargers.

Transaction costs—The administrative costs incurred by dischargers and regulatory agencies to administer and/or participate in water-quality-trading programs.

Trading ratio—A trading ratio is a requirement that credits be exchanged at other than a one-to-one ratio. There are various types of trading ratios serving different purposes.

Wasteload allocation—The pollutant load allocated to point-source dischargers in the total maximum daily load process.

Wastewater treatment plant—A point-source discharger requiring a National Pollutant Discharge Elimination System permit. A wastewater treatment plant can be publicly or privately owned and operated and can treat and discharge domestic or industrial wastewater, or both.

Water-quality-based effluent limits—Limits established by state regulatory agency for point-source dischargers into a water body, based on a determination of the wasteload that should be allocated to such dischargers.

APPENDICES

A Synopsis of Michigan's Water-Quality-Trading Regulations

Introduction

This appendix presents a detailed synopsis of the Michigan Rules (Michigan Department of Environmental Quality, 2002). It is nearly (but not completely) comprehensive. It is intended to give the reader an understanding of the strategy and content of the rules, without including every detail or regulatory boilerplate. Because it is not complete and is simply the author's interpretation and restating of the rules, anyone wishing to pursue water-quality trading in Michigan should consult the actual rules and not rely solely on this summary. The reference at the end of this appendix contains an internet link to the rules.

Rule 1—Definitions

Rule 2—Purpose

The purpose of the rules is to improve water quality and optimize the costs of achieving and maintaining water-quality standards. The rules create economic incentives for the following:

- Voluntary nonpoint-source-load reductions;
- Point-source discharge reductions beyond those required by the Clean Water Act (CWA, 1972);
- Implementation of pollution prevention programs;
- Wetland restoration and creation; and
- Development of emerging pollution control technologies.

The incentives facilitate the implementation of the following CWA requirements:

- Total maximum daily loads (TMDLs),
- Urban stormwater control programs, and
- Nonpoint-source-management practices.

The rules provide incentives for the development of new and more accurate and reliable quantification protocols and procedures.

They also provide greater flexibility through community-based, nonregulatory, and performance-driven, watershed-management planning.

Rule 3—Applicability

The rules apply to all persons and sources that participate in water-quality trading.

They apply to the generation, registration, use, banking, and trading of credits and all trading activities.

Rule 4—General Requirements

The generation, use, and trading of credits shall occur within the same receiving water or watershed designated under this part.

Credits shall be generated before or contemporaneously with the time they are used or traded.

The generation, use, and trading of credits and all trading activities shall be consistent with the following, if applicable:

- Total maximum daily load,
- Remedial action plan,

- ■ Lakewide management plan, or
- ■ Watershed-management plan approved by the Department of Environmental Quality (DEQ).

Credits used to comply with an effluent limitation established to achieve or maintain water-quality standards in a stream or lake with a retention time of less than one year shall be generated during the same time period for which the effluent limitation applies (e.g., daily, weekly, monthly, or annually).

Rule 5—Prohibitions and Restrictions

The use of credits that would cause a violation of water-quality standards is prohibited.

Credits generated in one watershed shall not be used or traded in a different watershed.

Credits generated in a non-attainment area can be used in an attainment area, if it is in a watershed designated in an approved watershed-management plan.

Trading activities for bioaccumulative chemicals of concern are prohibited (chemicals are listed).

Trading cannot be used to comply with technology-based limits.

Pretreatment trading is not covered by the rules.

Use of banked credits must meet the following conditions:

- ■ Must be preapproved;
- ■ Can be used to comply with the 1 mg/L total phosphorus requirement of R 323.1060(1); and
- ■ Can be used to comply with water-quality based effluent limits by a source that discharges to a lake or other water body with a retention time of more than one year.

Rule 6—Eligibility Requirements for Generation of Load Reductions and Credits

Point Source and Nonpoint Sources

The discharge or load reductions shall be real, surplus, and quantifiable.

The control devices or management practices that have been installed or implemented have been fully and properly maintained, from the time they were established, and remain so for the time they are registered to generate credits.

Included is a list of eligible ways to generate credits plus a catch-all way called "other pollution controls or management practices approved by the Department".

Discharge or load reductions required to achieve compliance with a technology-based effluent limitation established by an applicable requirement are not eligible to generate credits.

Generators and users of credits must discharge to the same receiving water.

If monitoring, recordkeeping, or reporting requirements are violated, credits will not be eligible.

The elimination of manure runoff or discharge shall be eligible only for the first five years after the effective date of the regulations.

Generally accepted agricultural management practices cannot be used to generate credits.

Nonpoint-source-load reductions funded with Natural Resource Conservation Service (NRCS) cost-share programs are eligible to generate credits in direct proportion to the local match percentage and any contribution greater than the local match required by these programs.

Nonpoint-source-load reductions, which result from implementation of programs funded by 1998 PA 288, MCL 324.19601 et seq. and §319 of the CWA (1972), are not eligible to generate credits.

Municipalities can generate credits by installing controls or implementing management practices under publicly funded projects or programs.

Rule 7—Nutrient Trading, Contemporaneous Upstream Reduction Requirements, and Credit Use

Nutrient trading may occur in an attainment area or impaired water where a TMDL has not been established and a watershed-management plan has not been approved, if either of the following conditions is met:

- There is a contemporaneous upstream generation of credits to compensate for a use of credits to comply with a water-quality-based effluent limitation or other requirement; or

- The source using credits to comply with a water-quality-based-effluent limitation or other requirement discharges to the same receiving water or watershed, either upstream or downstream of the source, which generates the credits, and both of the following conditions are met:

- The generation of credits is contemporaneous with the use of credits; or

- The sources, which generate and use credits, are upstream of the site of concern in the receiving water.

The use of nutrient credits is limited to a 20% increase above the "non-trading" level specified in a National Pollutant Discharge Elimination System (NPDES) permit, unless authorized by special conditions in the permit or a formal permit modification. The use of the 20% increase through trading must be authorized in the permit.

Rule 8—Nutrient Trading in Areas for Which a Total Maximum Daily Load or a Watershed-Management Plan Has Been Established

The rules refer to this as "closed" trading. It can occur if

■ A TMDL or watershed-management plan has been adopted, and

■ The provider and user of credits are located in non-attainment area addressed by the TMDL or watershed-management plan.

The TMDL or watershed-management plan allocations form the baselines for trading, and trading must be consistent with the TMDL or watershed-management plan and their allocations.

Rule 9—Other Types of Trading, Trading of Pollutants Other than Nutrients, Intra-Plant Trading, Cross-Pollutant Trading, and Trading under a Remedial Action or Lakewide Management Plan

Types of water-quality-based trades, other than nutrients, may be approved. These could include intra-plant trading or cross-pollutant trading. Such trades must be authorized in NPDES permits.

Rule 10—General Baseline Requirements

Baselines must be established using the most accurate and reliable data available.

For all sources except stormwater, data from the three-year period preceding the change made to generate a discharge or load reduction shall be used to characterize the baseline.

The baseline for stormwater sources regulated under an NPDES permit, for which a numerical effluent limitation has not been established, shall be the pollutant-specific loading achieved through implementation of management practices specified in or approved under a NPDES permit, at the time a change is made to generate a discharge or load reduction.

Baselines for agricultural, industrial, urban, and residential stormwater runoff shall be calculated by using the meteorological information and precipitation data for a 10-year period or the period-of-record, whichever is longer.

Rule 11—Baseline for Point Sources other than Stormwater, Reduced Discharge Level, and Generation of Discharge Reductions and Credits

The point-source baseline shall be the actual or allowed discharge level that complies with the most protective of any of the following:

- ■ A water-quality-based effluent limitation,
- ■ A cap and wasteload allocation specified under a TMDL,
- ■ A cap and wasteload allocation specified in a watershed-management plan, or
- ■ A cap and wasteload allocation specified in a remedial action plan or lakewide management plan.

Margins of safety achieved in practice shall be maintained by using the actual discharge flows and concentrations to calculate the baseline under subrule (3) of this rule.

The formula

Daily load = Flow × Concentration × 8.346

shall be used to calculate baseline discharges and discharge reductions.

Rule 12—Baseline for Stormwater Regulated under a National Pollutant Discharge Elimination System Permit, Reduced Discharge or Loading Level, and Generation of Discharge or Load Reductions and Credits

The baseline shall be the numerical effluent limit or the pollutant-specific loading achieved after implementation of NPDES permit requirements.

The baseline, reduced discharge level, generation of discharge reductions, and credits for stormwater sources, with numerical effluent limits or management practices specified by an NPDES permit, shall be calculated using the formula specified in Rule 11.

Monitoring data and actual measurements of load reductions will be used where possible.

Rule 13—Baseline for Unpermitted Nonpoint Sources of Stormwater Runoff other than Agriculture, Reduced Loading Level, and Generation of Load Reductions and Credits

The stormwater runoff baseline shall be either of the following:

■ For nonpoint sources that are not subject to an applicable requirement, the pollutant-specific loading associated with existing land uses and management practices, if any; or

- For nonpoint sources that are subject to an applicable requirement, the most protective of the following:
    - A cap and loading allocation specified in a TMDL,
    - A cap and loading allocation or the management practices specified in a watershed-management plan, or
    - A cap and loading allocation or the management practices specified in a remedial action plan or lakewide management plan.

Pollutant loads from various land uses are to be calculated using "Event Mean Concentrations" contained in Rule 13. Stormwater runoff from each land use is to be calculated using a formula specified in Rule 13. Pollutant removal rates for stormwater retention are specified in the rule. Pollutant loading is then calculated from these results. These methods are used to calculate baselines, reduced loading levels, load reductions generated, and credits.

Actual data may be used where available.

Rule 14—Agricultural Nonpoint-Source Baseline, Reduced Loading Level, and Generation of Load Reductions and Credits

The agricultural baseline shall be the most protective of the following:

- The pollutant-specific loading from existing agricultural operations that are not subject to an applicable requirement,

- The pollutant-specific loading achieved after implementation of management practices established by an applicable requirement,

- A pollutant-specific cap and loading allocation specified in a watershed-management plan, or

- A pollutant-specific cap and loading allocation specified in a remedial action plan or lakewide management plan.

Agricultural baselines must be established by a plan prepared by a planner certified by NRCS.

The plan must include the following:

■ Documentation of existing agricultural operations and management practices;

■ Quantification of the pollutant-specific loading from existing practices;

■ Identification of operational changes and management practices, which may be implemented to reduce loadings; and

■ Quantification of the pollutant-specific load reductions from each operational change and management practice recommended in the plan.

The baseline and pollutant-specific reduced loading level for each operational change and management practice recommended in the plan shall be established by one of the following methods and procedures:

■ For sediment, sediment-born nutrients, and concentrated animal feedlot runoff, "pollutants controlled calculation and documentation" (Michigan Department of Environmental Quality, 1999);

■ For commercial fertilizer application and manure management, methods and procedures approved by DEQ on a case-by-case basis; or

■ Alternate methods and procedures or models provided electronically by DEQ may be used for sediment, sediment-born nutrients, concentrated animal feedlot runoff, commercial fertilizer application, and manure management, when they become available.

The same methods and procedures shall be used to calculate the baseline, reduced loading level, load reductions generated, and credits.

Rule 15—Streambank Erosion Nonpoint-Source Baseline, Reduced Loading Level, and Generation of Load Reductions and Credits

The baseline for streambank erosion sources shall be one of the following, whichever is applicable and most protective:

- The pollutant-specific loading from existing sources that are not already subject to a requirement,

- The pollutant-specific loading achieved after implementation of management practices established by an existing requirement,

- A pollutant-specific cap and loading allocation specified in a watershed-management plan, or

- A pollutant-specific cap and loading allocation specified in a remedial action plan or lakewide management plan.

The baseline and pollutant-specific reduced loading level shall be established by the most conservative of the following methods:

- The use of aerial photographs;

- Calculated lateral recession rates;

- Gully erosion estimates at one-half of the amount calculated, in accordance with the U.S. Department of Agriculture field office technical guide for Michigan; or

- Other methods or procedures approved by DEQ.

The same methods and procedures shall be used to calculate the baseline, reduced loading level, load reductions generated, and credits.

Rule 16—Water-Quality Contribution and Uncertainty

For point sources other than stormwater, 10% of the reductions generated will be contributed to DEQ to address uncertainty and provide a net water-quality benefit.

This is a "one-time" contribution (simply meaning it is paid, in full, at the time the trade is initiated, even if the trade covers multiple years).

For stormwater point sources, the contribution is 50%, unless a lesser percentage is granted by DEQ.

Rule 17—Discount Factors Applied to the Use of Credits

If a lake, impoundment, or wetland is located between the user of credits and an upstream supplier, then the user shall acquire 10% more credits than needed.

A source using credits in a non-attainment area with no TMDL shall obtain 10% more credits than needed to comply with existing requirements.

The DEQ may require other discount factors.

Rule 18—Nutrient Discharge and Load Reductions and Early Reductions and Credit Life

Banked credits for phosphorus and nitrogen, which are entered in the trading registry, may be used or traded for a period of five calendar years after the year of generation, subject to the prohibitions, restrictions, and conditions established in this part.

Phosphorus and nitrogen reductions achieved in advance of a reduction requirement may be banked.

Banked credits that are not used will be retired to achieve a water-quality benefit.

Rule 19—Notification Requirements for Generation of Discharge, Load Reductions, and Registration of Credits

Anyone wanting to sell credits shall register them with DEQ.

The notification to DEQ must include the following:

- The name and location of the source generating the credits;
- Documentation of baseline, load reductions, and credits generated, by watershed;
- Methods used to generate the load reductions;
- Date that the load reduction will take effect and the period of time that the reduction will remain in effect; and
- Quantification and monitoring methods.

The responsible individual submitting the notification must certify that

- The information is true, accurate, and complete;
- The load reductions are real, surplus, and quantifiable, and will be generated in the appropriate time period; and
- The load reductions have not been used elsewhere as credits.

The notice shall be submitted electronically or by certified mail to DEQ.

The DEQ shall review the notice and make a determination of completeness and consistency and provide a written response to the person submitting the notice within 30 days. This does not constitute DEQ certification that the credits are real, surplus, or quantifiable, only that the submission is complete and consistent with the regulations.

The DEQ shall enter the necessary data in the water-quality-trading registry within five business days of the determination of completeness and consistency. The information is then available to the public.

The DEQ shall explain any determination of incompleteness or inconsistency. Following such a determination, the applicant may submit a corrected or revised notice and certification.

Once DEQ issues a determination of completeness, the methods used to generate the credits become legally enforceable requirements.

Issuance of a notice of credit generation by DEQ shall constitute notice that a point source is subject to alternate NPDES permit limits for the period specified in the notice.

Point sources (except stormwater) shall report the baseline, quantity of discharge reductions, and credits generated on their monthly discharge-monitoring reports.

Nonpoint and stormwater sources with NPDES permits (without numerical limitations) shall submit quarterly reports to DEQ that include the following:

- Name and location of the site;

- Pollutants controlled;

- Control devices installed or management practices implemented and date completed;

- The lineal feet or acres for which controls or management practices have been completed; and

- A calculation of the quantity of each pollutant controlled, using the same methods and procedures used to determine the baseline, load reductions, and credits.

Rule 20—Notification Requirements for the Registration, Use, and Trading of Credits

Anyone wanting to use credits shall provide prior notice to DEQ. The notification to DEQ must include the following:

- ■ Name of responsible individual and name and location of usage;

- ■ Documentation of effluent or other limits, the number of credits to be used to comply, and the new "credit-adjusted" discharge limit;

- ■ Quantity of credits used, by watershed;

- ■ A description of the source, process, or operation at which the credits are to be used;

- ■ Methods and procedures used to quantify loading and determine compliance with all requirements;

- ■ Dates of the use of credits; and

- ■ A copy of the notice of generation of credits, filed by the producer of the credits.

The responsible individual submitting the notification must certify that

- ■ The information is true, accurate, and complete; and

- ■ The source, process, or operation shall be operated in compliance with all requirements, including those for use of credits.

The notice shall be submitted electronically or by certified mail to DEQ.

The DEQ shall enter the proposed notice of use in the registry within three business days.

The DEQ shall review the notice and make a determination of completeness and consistency within 30 days and provide a written response to the notifier.

The DEQ shall enter the information in the water-quality-trading registry within five business days. The information is then available to the public.

The DEQ will not issue a notice for a proposed use of credits that it determines would violate water-quality standards.

The DEQ will explain any determination of incompleteness or inconsistency.

The methods used and operational changes made to use the credits shall become legally enforceable operating requirements, effective on the date DEQ issues a notice of completeness.

Users of credits must notify DEQ of the price paid for the credits.

Following the end of the credit use period, the user has 60 days to notify DEQ of any unused credits.

Issuance of a notice for use of credits by DEQ constitutes notice that a point source is subject to alternate NPDES permit limits.

Point sources must report to DEQ, in their monthly discharge-monitoring reports, the number of credits used. Nonpoint sources must submit an annual report.

Rule 21—Water-Quality Trading Registry

The DEQ will maintain a trading registry for the following purposes:

■ Registering discharge and load reductions;

■ Registering and tracking the generation use and trading of credits;

■ Registering the credits that are contributed to the state as water-quality contributions; and

■ Providing the public access to the data.

The registry will contain all of the information submitted under Rules 19 and 20.

The registry shall be updated daily and made available to the public through an electronic bulletin board.

Rule 22—Delineation of Watersheds for Purposes of Water-Quality Trading

Rule 23—Watershed-Management Plans for Water-Quality Trading, Submittal, and Approval

Trading may occur under the following plans:

- ■ Total maximum daily loads;
- ■ Remedial action plans;
- ■ Lakewide management plans;
- ■ Watershed-management plans (319 plans);
- ■ Watershed-based, stormwater-management program under a NPDES permit;
- ■ Watershed-based, stormwater-management program under a voluntary national permit; or
- ■ Nonpoint-source, watershed-management plans developed under a Clean Michigan Initiative grant.

The plans must describe the role of trading in accomplishing the plan's goals.

Comprehensive watershed-management plans using trading for any of the following purposes may be submitted for DEQ approval:

- ■ Improving water quality and enhancing aquatic habitat;
- ■ Reestablishing or creating wetlands or floodplains;
- ■ Encouraging environmentally sound land-use practices;
- ■ Accommodating growth and economic development;
- ■ Creating nature conservancies, parks, and natural areas; or
- ■ Other.

Rule 24—Program Evaluation

To assess the environmental and economic performance of the trading program, DEQ shall evaluate the program after three years and then conduct watershed specific evaluations every five years, concurrent with ambient monitoring and NPDES permitting cycles.

The evaluations shall include the following information:

- Identification of watersheds where trading has occurred. The identification shall include the following:
  - Trading area,
  - Number and type of point and nonpoint sources, and
  - Water-quality status.
- Ambient monitoring conducted by DEQ or others to quantify actual nonpoint-source-load reductions and assess water quality;
- Type and number of trades, by pollutant;
- Quantity of credits traded;
- Quantity of credits that have been retired;
- Comparison of costs (trading versus no trading, if information is available);
- Price paid for credits, by pollutant;
- Costs incurred by DEQ; and
- Transaction costs incurred by trading participants, if information is available.

The DEQ shall use this information to determine if the trading program has

- Been consistent with achieving and maintaining water-quality standards;
- Achieved a net reduction in the loadings of pollutants from trading partners;

- Achieved voluntary and early reductions of pollutant discharges and loadings;

- Resulted in the development of emerging pollution-control technology or new or improved methods and procedures for the quantification of loads;

- Caused any localized adverse effects to the public health, safety, welfare, or environment; and

- Had sufficient accountability and compliance.

The DEQ shall propose any necessary program modifications.

The program evaluation shall be made available for public comment. Following public comment, necessary revisions may be made to the program.

Rule 25—Compliance and Enforcement

Any source that uses credits is solely responsible for compliance with all of its effluent limits, notwithstanding another person's liability, negligence, or false representation.

A source that registers credits that are traded shall be strictly liable for assuring that the reductions are real, surplus, quantifiable, and equal to the quantity of credits that are registered.

A source may notify DEQ that the quantity of discharge or load reductions actually generated or the quantity of credits used or traded are not real, surplus, quantifiable, or are insufficient.

A source providing such notice shall be provided a reconciliation period of not more than 30 days to resolve the insufficient reductions or credits, if the following conditions are met:

- The notice is submitted to DEQ within seven days of the discovery of the problem, and

- The notice includes all of the following information:

  - An explanation of how and when the deficiencies were discovered;

■ Corrective actions taken or planned, and their schedule;

■ A revised notice and certification of load reduction or credit use; and

■ Certification that the information is true, accurate, and complete.

The person submitting the notice shall also do either of the following, as applicable:

■ If insufficient credits have been traded, then the source submitting the notice shall implement and register load reductions or obtain credits from another source to compensate for the insufficient credits; or

■ If the credits had not been used or traded, then the source who registered them shall file a revised notice of generation or request that DEQ remove the credits from the registry.

If DEQ discovers that registered credits are not valid or more credits have been traded than were produced, then the source that generated the credits must generate or obtain triple the amount of invalid credits and donate them to DEQ for retirement.

The DEQ may also take appropriate enforcement actions under the CWA (1972) or Michigan law. In any such action, the source generating and registering the credits has the burden of proof to show that the credits are valid.

A source that uses credits later determined to be invalid shall have a reconciliation period of 90 days from the date of discovery to produce or acquire credits to compensate for the invalid ones.

A source that knows or should have known that credits were not valid shall not be entitled to the reconciliation period.

Rule 26—Availability of Documents

Rule 27—Availability of Federal Regulations

References

Clean Water Act (1972) U.S. Code, Section 1251–1387, Title 33.

Michigan Department of Environmental Quality (1999) *Pollutants Controlled and Calculation and Documentation for Section 319 Watersheds Training Manual*; Nonpoint Source Unit, Surface Water Quality Division, Michigan Department of Environmental Quality, Lansing, Michigan; http://www.deq.state.mi.us/documents/ deq-swq-nps-POLCNTRL.pdf (accessed January 28, 2005).

Michigan Department of Environmental Quality (2002) *Rule Part 30: Water Quality Trading* (effective November 22, 2002); Michigan Department of Environmental Quality: Surface Water Quality Division: Lansing, Michigan; http://www.state.mi.us/orr/emi/ arcrules.asp?type=Numeric&id=1999&subId=1999%2D036+EQ &subCat=Admincode (accessed April 16, 2004).

Best Management Practice List for the Lower Boise River Pollution Trading Program

BEST MANAGEMENT PRACTICE (BMP) LIST FOR THE LOWER BOISE RIVER POLLUTION TRADING PROGRAM

THE LOWER BOISE RIVER POLLUTION TRADING PROJECT

This Pollution Trading project has been established and supported by many agencies and local interests to assist the point and nonpoint phosphorus sources in reducing their phosphorus loads and implementation costs in meeting a Total Maximum Daily Load (TMDL) at the mouth of the Boise River near Parma, Idaho. A "trading market" should enable point and nonpoint sources reductions to be achieved at lesser costs.

The trading that occurs between point and nonpoint sources will be due largely to high point source reduction costs. The point sources that cannot immediately meet their permitted discharges would be permitted to discharge in excess of their permit as long as there is an equal reduction at another point or nonpoint source location. In-stream water quality problems due to discharges in excess of what is permitted will not be allowed under this trading program. Water quality improvements are still to be achieved, regardless of the activity within the trading program.

DOCUMENT PURPOSE

Selected nonpoint source BMPs can be used to offset a point source's discharge, in which are described here. The procedure for generating credits, as well as other trading program requirements, are described as well. This document will be updated periodically and new BMPs added to the list of those currently eligible for trading.

CALCULATED AND MEASURED PHOSPHORUS CREDITS

To offset a given amount of phosphorus at one location from a point source, there must be an equal and beneficial reduction from another point or nonpoint source location. The term "credit" has been established to represent that equalized portion of phosphorus considered in the trading market. The reduction is calculated or measured in pounds of phosphorus, determined by one of two methods. These reductions are then converted to credits for trading purposes.

To estimate what a BMP's capability is in reducing phosphorus losses, local sampling data is needed in order to make that estimate. Where there is adequate data for a specific BMP's reduction capability, a calculation can be made with fair certainty of it actually occurring. Where data is limited, "measuring" for phosphorus removal is necessary. For this trading program, participants will use either the calculated or measured approach to generate credits. The calculated approach will utilize exiting data to estimate an average reduction for a particular BMP, with a slight discount in its effectiveness due to potential uncertainty in the data and other management factors. For measured credits, grab samples will be taken during the BMP's operation to quantify the actual reductions. An inflow and outflow condition will be necessary to sample a BMP.

GENERAL BEST MANAGEMENT PRACTICE (BMP) REQUIREMENTS FOR THE POLLUTION TRADING PROJECT

Agricultural landowners participating in the pollution trading program are highly encouraged to develop a conservation plan with one of two Soil Conservation Districts (SCD). The Ada Soil Conservation District resides at 132 SW 5th Ave., Meridian, ID 83642 (208-888-1890 x3) along with the Natural Resource Conservation Service (NRCS), the Soil Conservation Commission (SCC), and the Farm Services Agency (FSA). Ada county participants will utilize this office for technical and trading program assistance. For Canyon county participants, the Canyon Soil Conservation District is located at 2208 E. Chicago St. Caldwell, ID 83605 (208-454-8684), which also includes NRCS, SCC, and FSA.

The conservation plans are cooperatively developed among the landowner, NRCS and the SCC. These conservation plans are developed to address existing natural resource concerns as well as meeting the landowner's objectives. Through the conservation planning process, BMP installation and other planned activities are evaluated to ensure that they do not have significant negative impacts on natural resources and other landowners.

The BMPs typically used to address water quality concerns are listed in the Agricultural Pollution Abatement Plan (APAP), which is kept at the SCC. BMPs originate in the USDA-NRCS National Handbook of Conservation Practices (NHCP, 2000), which can be found in either of the SCD offices.

Upon installation, after being incorporated into this document, it is to be certified as installed according to NRCS and this document's criteria, as well as meet any applicable local, state, and federal laws and regulations. Upon certification and at the start of BMP operation, credit generation can begin. Most agricultural BMPs within the Lower Boise River watershed will provide reductions primarily within the irrigation season as designed and operated. All BMPs are to function according to the appropriate criteria throughout their operating period.

All BMPs are to be inspected after installation or application, prior to their seasonal period operation. Some BMPs will require a greater number of inspections as outlined in the monitoring section.

CURRENT ELIGIBLE BMPS FOR TRADING

The program eligible BMPs are listed in Table 1, which are also discussed in Carter 2002. The NRCS practice code and typical lifespan are included here.

Table 1. BMPs Currently Eligible for Trading.

BMP	NRCS Code[1]	Lifespan
Sediment basins	350	20 years
Filter strips	393	1 season
Underground outlet	620	20 years
Straw in furrows	484	1 season
Crop sequencing	328, 329	1 season
Polyacrylamide	450	1 irrigation
Sprinkler Irrigation	442	15 years
Microirrigation	441	10 years
Tailwater Recovery	447	15 years
Surge Irrigation	430HH	15 years
Nutrient Management	590	1 year
Constructed Wetland	656	15 years

[1] Refer to http://id.nrcs.usda.gov/practices.htm
Additional components for the BMP may incorporate other practice codes.

BMP EFFICIENCY AND UNCERTAINTY DISCOUNTS

Listed in Table 2 are the effectiveness and uncertainty discounts for the currently eligible types, field, farm, and watershed scale. The sediment basin is categorized into 3 types, which, are due to differences in the size of treatment area and duration of flow in the basins.

Nutrient management does not have a phosphorus reduction efficiency due to numerous complexities. This practice is, however, a necessary long-term practice that will benefit water quality if applied properly. Though this practice does not have an efficiency associated with it, it is a valuable BMP for this trading program and will be marketable in relation to other applied BMPs. If nutrient management is applied in addition to other eligible BMPs, the uncertainty factor for those other BMPs will reduced by 50%, thereby, increasing their market value.

Table 2: BMP Effectiveness and Uncertainty Discounts

BMP	Effectiveness	Uncertainty[1]
Polyacrylamide	95%	10%
Filter Strip	55%	15%
Sprinkler	100%	10%
Microirrigation	100%	2%
Tailwater Recovery	100%	5%
Mulching	90%	20%
Crop sequencing	90%	10%
Sediment Basin Field scale	80%	10%
Sediment Basin (farm scale)	75%	10%
Sediment Basin (watershed scale)	65%[4]	15%[4]
Underground Outlet	85% (65%)[2]	15% (25%)[2]
Surge Irrigation	50%	5%
Nutrient Management	NA[3]	NA[3]
Constructed Wetland (farm scale)	90%	5%
Constructed Wetland (watershed scale)	NA[4]	NA[4]

[1] This is to be subtracted from the efficiency.
[2] This BMP's effectiveness drops after 2 years.
[3] Data unavailable for efficiency estimate. If applied with other eligible BMPs, their uncertainty discounts will be reduced by 50%.
[4] Not recommended for calculated credit.

BMP MONITORING: EVALUATION AND MEASUREMENT REQUIREMENTS

To ensure that a BMP is operating properly and actually reducing phosphorus losses, an evaluation is necessary. An evaluation will consist of at least 1 annual field inspection to ensure proper application and operation. Table 3 provides the minimum inspections needed for each BMP, and provides a minimal level of measurement requirements, though not applicable to all BMPs.

Some BMPs do not allow for true "inflow-outflow" comparisons utilizing flow and nutrient measurements, therefore it is not recommended for measurement. Also, a measurable BMP's inflow conditions only represent the instantaneous condition, not reflective of the 1996 baseline condition. In essence, these instantaneous measurements would provide a pretreatment load different than that of the baseline average load, misrepresenting the average 1996 loads. Therefore, no measurements will be allowed for field-scale BMPs to generate credits.

Watershed-scale BMPs, such as the sediment basin and constructed wetlands, where they are not easily calculated, will be measured to generate credits. The schedule for measurements will be set within the buyer-seller contracts for specific watershed-scale BMPs.

Table 3. BMP Evaluation Requirements

BMP	Evaluation
Sediment basin - field scale	before & middle of all irrigations
Sediment basin - farm scale	before & middle of all irrigations
Sediment basin - watershed scale	before & middle of season of use
Filter strips	before & middle of all irrigations
Underground outlet	before & middle of all irrigations
Straw in furrows	before & middle of all irrigations
Crop sequencing	before & middle of all irrigations
Polyacrylamide	evaluate 2 irrigations & review application records
Sprinkler Irrigation,	evaluate 1 irrigation
Microirrigation	evaluate 1 irrigation
Tailwater Recovery	before irrigations & evaluate 1 irrigation
Surge Irrigation	evaluate 1 irrigation
Nutrient Management	evaluate records annually
Constructed wetland	before & middle of season of use

CREDIT PRODUCTION METHOD

Calculated Credits

To calculate a total phosphorus credit, a reduction estimate is determined prior to the sale of the credits, utilizing BMP effectiveness data and other applicable factors.

In the case of calculated credits, specifically to a cropland field, the phosphorus losses in 1996 (TMDL baseline) must be estimated. The Surface Irrigation Soil Loss (SISL) tool is currently the most accurate and simple method available for the program area to estimate soil losses from surface irrigated croplands. SISL losses are then converted to phosphorus losses by multiplying tons soil loss by 2, which provides pounds of phosphorus. Typically, there is on average, 2 pounds of phosphorus loss per ton of soil loss within the program area. This tool is described in USDA-NRCS Agronomy Technical Note No. 32.

There is a great amount of variability in soil and phosphorus loss from one year to the next because of crop rotations, as the SISL shows when used according to its design. This variability would cause a great deal of fluctuation from year-to-year in credits generated from one field. This fluctuation may is not greatly desired in a trading program. Also, because there does not exist data for all fields within the program area for 1996, the crop specific SISL estimate cannot be derived for a number of fields.

An average subwatershed Base Soil Loss (BSL), a necessary factor in SISL, has been determined for each the major Lower Boise River subwatersheds (Table 4). Numerous field crop records from 1996 were evaluated to establish baseline 1996 soil losses with SISL. By utilizing the average subwatershed BSL, crop rotations will have no effect on credit calculation because the pretreatment load of 1996 will not change. A change in credits will only be due to switching from one BMP to another.

Where the SISL-BSL represents seasonal sediment losses, monthly losses may be estimated utilizing numerous irrigation records, which can be used to provide an average number of irrigations per month. Another critical factor to be considered in determining an average sediment and phosphorus loss on a monthly basis, is the percent soil loss of total per irrigation. The first three irrigations typically produce the majority of the annual sediment loss, whereas, with each additional irrigation, less erosion takes place due to increasing soil stability and some crop foliage protection where it lies within the furrow later in the growing season.

Table 4. SISL BSL (tons/ac/yr soil loss[1]) per Subwatershed

Slope of field	<1%		1-1.9%		2-2.9%		>3%	
Drain/Field length	660	1320	660	1320	660	1320	660	1320
Eagle Drain	2.0	1.6	7.3	5.8	15.5	12.4	25.2	20.2
Thurman Drain[2]	NA	NA	NA	NA	NA	NA	NA	NA
Fifteenmile	1.6	1.3	5.8	4.6	12.5	10.0	21.0	16.8
Mill Slough	2.0	1.6	7.3	5.8	15.5	12.4	25.2	20.2
Willow Creek	1.9	1.5	6.8	5.5	14.7	11.7	24.0	19.2
Mason Slough	2.0	1.6	7.3	5.8	15.5	12.4	25.2	20.2
Mason Creek	1.7	1.4	6.4	5.1	14.1	11.2	23.7	18.9
East Hartley	2.0	1.6	7.3	5.8	15.7	12.5	25.6	20.5
West Hartley	2.0	1.6	7.3	5.8	15.7	12.5	25.6	20.5
Indian Creek	1.9	1.5	6.9	5.5	14.9	11.9	24.7	19.8
Conway Gulch	2.0	1.6	7.3	5.8	15.7	12.5	25.6	20.5
Dixie Drain	1.7	1.4	6.4	5.1	13.9	11.1	23.0	18.4
Boise River	2.0	1.6	7.3	5.8	15.5	12.4	25.2	20.2

[1] Multiple BSL by 2 to obtain pounds of phosphorus
[2] Thurman drain currently does not have any cropland fields within it drainage area.

Based on numerous irrigation records and local input, average number of irrigations per crop type per month was established, then one average for all crops per month. The average number of irrigations per month is shown in Table 5.

Table 5. Average Number of Irrigations per month, based on a 181-day irrigation season.

Month	Irrigations	Days/month
April	0.4	15
May	1.2	31
June	2.4	30
July	3.0	31
August	1.9	30
September	0.5	31
October	0.2	15
Total	9.5	181

The average number of irrigations per month was not rounded to the whole number because it would exclude any irrigation that does occur in April and October. The irrigation season is assumed to start on start on April 15 and end October 15, providing a 181 irrigation day season.

Based on numerous runoff studies on surface irrigated cropland, percent soil loss per irrigation was determined. These percent losses per irrigation were then lined up with the average 9-10 irrigations per season to estimate average percent loss per irrigation (Figure 1).

Figure 1. Average Percent Soil Loss per Irrigation per Total Season Loss

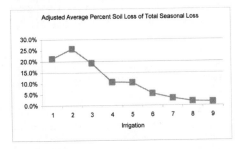

Table 6 shows the percent loss per month, which was derived from the average irrigations per month (Table 5) and percent loss per the 9-10 irrigations per season (Figure 1).

Table 6. Percent Soil Loss per Month

Month	Percent Loss
April	8.5%
May	28.1%
June	39.9%
July	19.4%
August	3.6%
September	0.4%
October	0.1%

Recent water quality samples taken throughout the Lower Boise River tributaries reflect similar loss characteristics, where the months of May, June, and July show the largest in-stream sediment loads. Once the seasonal SISL losses are determined, which represents the pretreatment load, a monthly estimate can be estimated with the values from Table 6.

River Location Ratios

Upon establishing a monthly or irrigation season phosphorus reductions, with a BMP applied,

pounds reduced are to be converted into "Parma Pounds" or credits. The current adopted method utilizes a simple mathematical calculation to convert pounds into credits. The amount of phosphorus retained by a BMP on a field within a subwatershed does equal the amount of phosphorus reduced at the mouth of the drainage. There are River Location Ratios (DEQ, 2000) that attempt to account for the river's phosphorus transmission losses and are set at the various locations within the river system, primarily at the mouths of the major tributaries, as shown in Table 7. Those river adjacent lands that impact the river directly will receive the next downstream tributary river location ratio.

Table 7. River Location Ratios

Subwatershed	River Location Ratio
Eagle Drain	0.63
Thurman Drain	0.51
Fifteenmile Creek	0.75
Mill Slough	0.75
Willow Creek	0.75
Mason Slough	0.75
Mason Creek	0.75
East Hartley Gulch[1]	0.80
West Hartley Gulch[1]	0.80
Indian Creek	0.89
Conway Gulch	0.95
Dixie Drain	0.96

[1] East & West Hartley Gulch merge before confluence at Boise River

Site Location Factors

Transmission losses may occur between the point where the reduction takes place and the subwatershed's channel due to wastewater being water reuse and natural sediment-phosphorus relationships. Canals may intercept wastewater runoff from fields, which may or mat not impact the drainage in which the field is located. The greater the travel distance and the chance of reuse, the less likely the total phosphorus amount lost at the field will reach the channel. Site Location Factors are developed to account for some of this transmission loss, shown in Table 8.

Table 8. Site Location Factors

Land runoff flows into a canal, likely to be reused by downstream canal users	0.6
Land runoff does not flow directly to a drain, but through or around other fields prior to entering a drain	0.8
Land runoff flows directly to a drain or stream through a culvert or ditch	1.0

Drainage Delivery ratios

Drainage Delivery Ratios were also developed to account for the phosphorus transmission losses in the subwatershed's main channels. Recent water quality samples collected from within some of these subwatersheds do show however, upstream to downstream, an increase in phosphorus concentrations. This increase in phosphorus concentration is likely due to increasing surface and ground water flows and phosphorus loads from increasing numbers of sources. Due to no available research data or locally developed transmission models, a simple linear calculation is made that represents this potential loss, which is:

(100 - distance in miles to mouth of the drain from the project's point of discharge on the drain)/100.

A measurement, in miles, is made from the mouth of the channel on the river to the point where the wastewater enters the channel. This measurement is to be made with the use of computer based Geographic Information Software (GIS).

Example Credit Calculation

The following is an example of the current method of calculating credits:

Given: 30 acre surface irrigated field to be converted to a sprinkler system, capable of eliminating all sedimentation loss (100% removal) but with a 10% uncertainty factor (subtracted from BMP efficiency). Average annual SISL load is determined to be 7.3 tons/acre (219 total) soil loss per irrigation season. Total annual phosphorus loss is

calculated to be 438 pounds (219 x 2 lbs/t). Assuming a 78% TMDL reduction requirement from all sources, 342 lbs is to be removed first, prior to trading and calculating credits. A total of 394 lbs is to be reduced by the sprinkler system (0.9 x 438). The Site Location Factor is 0.8, because there is potential but reuse, but not through a canal. The distance from the river to the entry point at the channel is 2.5 miles, which gives a 0.975 Drainage Delivery Ratio. The River Location Ratio is 0.75.

Credits (Parma Pounds) =
    438 lbs P
    x 0.90 (1.0 effective - 0.10 uncertainty)
    -342 lbs P (438 lbs P x TMDL 0.78)
    x 0.8 site location factor
    x 0.975 drainage delivery ratio
    x 0.75 river location ratio =
30 credits (Parma Pounds) for sale for one irrigation season.

By month:	*April*	*2.6*
	May	*8.4*
	June	*12.0*
	July	*5.8*
	August	*1.1*
	September	*0.1*
	October	*0.0*

INCLUSION OF NEW OR EXISTING PHOSPHORUS CONSERVATION MEASURES TO THE BMP LIST

There may be other conservation measures not specifically characterized within the NHCP or APAP that can reduce phosphorus losses from agricultural lands or treat wastewater. These conservation measures can be added to this list at any time, once they have been reviewed and approved by the BMP technical Committee potentially undergo a public review process to fulfill the trading program requirements.

Proposed conservation measures to be considered for the purpose of establishing credits not contained within this list are to be forwarded to the Idaho Soil Conservation Commission, BMP Technical Committee, Pollution Trading, P. O. Box 790, Boise, Idaho 83701 at (208) 332-8650.

REFERENCES

Carter, D. L. 2002. Proposed Best Management Practice (BMP) list and application criteria for the Lower Boise River Pollution Trading Demonstration Project, Unpublished report.

Idaho Department of Environmental Quality (DEQ). 6/7/2000. Lower Boise River pollution trading demonstration project, summary of participant recommendations for a trading framework. Unpublished document.

USDA - Natural Resources Conservation Service. 2000. Agronomy technical note 32, rev. 2. Predicting irrigation induced soil loss on surface irrigation cropland. Unpublished document. – USDA-NRCS, Boise Idaho.

APPENDIX C

Perpetual Conservation Easement

Perpetual Conservation Easement

This Conservation Easement, made this 30th day of April, 1997, between the City of New Ulm, Minnesota, a municipal corporation, herein referred to as "City," and Rahr Malting Co., a corporation under the laws of the State of Delaware, herein referred to as "Rahr."

Recitals

A.   Rahr desires to acquire a conservation easement on certain lands to establish vegetative cover and conservation practices, which will include the planting of native grasses and trees, in order to protect soil and water quality, and to enhance fish and wildlife habitat.

B.   The City is the owner of marginal lands, and/or drained or existing wetlands and/or crop land adjacent to these lands, and desires to convey a conservation easement to Rahr.

Agreement

NOW, THEREFORE, the City, for itself, its successors and assigns, in consideration of the sum of Fifty One Thousand Two Hundred Dollars ($51,200.00), does hereby convey and warrant to Rahr, its successors and assigns, forever, a perpetual conservation easement upon the following described land, herein referred to as the "easement area," situated in the County of Brown, State of Minnesota, to-wit:

Lot 1, Block 1, Riverview Addition
City of New Ulm, Brown County, Minnesota,

AND

Outlot C, Two Rivers Subdivision,
City of New Ulm, Brown County, Minnesota.

The easement area is subject to all easements, roadways, minerals and mineral rights of record. The City reserves all minerals and mineral rights in the easement area.

In addition, Rahr, for itself, its successors and assigns, and the City, for itself, its successors and assigns, agree as follows:

1. The City represents and warrants that the City, its successors and assigns, shall not place any foreign substances in or on the easement area that will cancel or invalidate Rahr's credits for nonpoint source trading and wetland conservation act credits.

2. Rahr shall establish and maintain a permanent vegetative cover on the easement area, including any necessary replanting thereof, and other conservation practices. The conservation practices shall include the planting of various native grasses and trees in accordance with the conservation plan for New Ulm properties contained in the memo dated April 15, 1997 from North American Wetland Engineering, P.A., attached hereto as EXHIBIT A. Rahr and the City agree that the conservation practices shall be such that they quality for credits for nonpoint source trading and wetland conservation act credits.

3. The City shall not appropriate water from any existing or restored wetlands within the easement area unless obtaining the prior written consent of Rahr and all necessary government permits.

4. The City shall not produce or allow to be produced agricultural crops on the easement area, except with the prior written approval of Rahr for wildlife or timber stand management purposes.

5. The City shall not graze or allow to be grazed any livestock on the easement area.

6. The City shall not place any materials, substances, or objects, nor erect or construct any type of structure, temporary or permanent, on the easement area without the prior

written approval of Rahr. Notwithstanding the previous sentence, the City specifically reserves the right to install utilities and associated, necessary structures and appurtenances on the easement area. The City may also make the easement area available for public enjoyment, e.g., nature trails, as long as the public enjoyment is not contrary to the purposes of the conservation easement expressed herein.

7. The City shall be responsible for weed control by complying with noxious weed control laws and emergency control of pests necessary to protect the public health on the easement area. The City's responsibility under this paragraph shall only arise after the State of Minnesota and the City have confirmed that a permanent vegetative cover has been established. Rahr shall re-establish any vegetative cover that is lost due to the City's actions to control noxious weeds or pests.

8. Except as provided in this agreement, the City shall not alter wildlife habitat, natural features, the vegetative cover, or other conservation practices on the easement area without the prior written approval of Rahr.

9. The City shall restore the easement area to its pre-existing condition after any lawful installation, repair, or improvement to any public utility system.

10. The City shall notify Rahr in writing of the names and addresses of new owners within thirty (30) days after conveyance of all or part of the title or interest in the land described herein.

11. The City shall pay when due all taxes and assessments, if any, that may be levied against the easement area.

12. The City shall undertake the protection of the easement area in accordance with the conditions set forth in this easement. Specifically, Rahr shall post the easement area prohibiting the use of motorized or other vehicles that would disrupt the vegetative cover, and the City shall enforce this prohibition. Rahr reserves the right to display its name upon any signs it posts within the easement area.

13. The City shall allow authorized agents of Rahr to enter upon the easement area for the purpose of inspection, management, and enforcement of this easement, together with the rights to ingress and egress to the easement area from a public road. Established access routes shall be used whenever practical.

14. Rahr shall retain rights to credits for nonpoint source trading and wetland conservation act credits.

15. Rahr shall be responsible for establishing the physical location of the easement area and shall confine its conservation practices and activities to the easement area.

16. Rahr and the City acknowledge that this conservation easement shall run with the land and shall be binding upon the parties, their successors, assigns, and tenants. In the event Rahr dissolves, ceases to operate its facilities located in Shakopee, Minnesota, and has no identifiable successors or assigns, the City shall no longer be bound by the provisions of this conservation easement.

17. This easement shall be enforceable by Rahr and/or by such other relief as may be authorized by law. Any ambiguities in this easement shall be construed in a manner which best serves the purpose of protecting soil, improving water quality, and enhancing fish and wildlife habitat.

Water-Quality-Trading Resources
Environmental Trading Network

The Environmental Trading Network (ETN) (Kalamazoo, Michigan) (originally called the Great Lakes Trading Network) began in 1998 to support the Kalamazoo River Phosphorus Trading Demonstration Project. The ETN quickly grew to be a national clearinghouse for water-quality trading information and the best opportunity available to network with people across the country involved with trading. Today, ETN is dedicated solely to the development and implementation of successful water-quality trading programs and other market-based strategies for achieving healthy sustainable ecosystems.

The ETN is accessible through its Web site (http://www.envtn.org), one of the best places to find information on water-quality trading. It contains trading news, information on trading programs, trading resources, and links to U.S. Environmental Protection Agency (U.S. EPA) trading documents and numerous other trading resources.

In addition, ETN hosts a monthly conference call to discuss the latest developments in trading across the country. Anyone can participate in these calls by adding their name to the contact list on the Web site at http://www.envtn.org.

U.S. EPA Watershed Trading Web Site

The U.S. EPA watershed trading Web page (http://www.epa.gov/owow/watershed/trading.htm) is one of the best sources available for information on trading. It contains frequently asked questions about trading, links to all U.S. EPA trading policy documents, links to trading projects and other trading information, information on trading conferences, and access to U.S. EPA's archives of trading-related documents.

Water Environment Research Foundation

The Water Environment Research Foundation (WERF) (Alexandria, Virginia) is a not-for-profit organization that

Seeks to promote the development and application of sound science to water-quality issues. WERF subscribers include municipal and regional water and wastewater utilities, industrial corporations, environmental engineering firms, and others that share a commitment to cost-effective water quality solutions that protect the environment and improve the quality of life for all.

The Water Environment Research Foundation funds water-quality research and publishes the research reports. The reports are available for purchase on the WERF website (http://www.werf.org). The research reports on water-quality trading that are currently available (as of April 19, 2004) include the following:

Bacon, E. (2002) *Nitrogen Credit Trading in Maryland: A Market Analysis for Establishing a Statewide Framework*; Water Environment Research Foundation: Alexandria, Virginia.

Baumgart, B. P.; Johnson, B. N.; Pinkham, J. R. (2000) *Phosphorus Credit Trading in the Fox-Wolf Basin: Exploring Legal, Economic, and Technical Issues;* Water Environment Research Foundation: Alexandria, Virginia.

Kieser, M. S. (2000) *Phosphorus Credit Trading in the Kalamazoo River Basin: Forging Nontraditional Partnerships*; Water Environment Research Foundation: Alexandria, Virginia.

Moore, R. E.; Overton, M.; Norwood, R. J.; DeRose, D. (2000) *Nitrogen Credit Trading in the Long Island Sound Watershed*; Water Environment Research Foundation: Alexandria, Virginia.

Paulson, C.; Vlier, J.; Fowler, A.; Sandquist, R.; Bacon, E. (2000) *Phosphorus Credit Trading in the Cherry Creek Basin: An Innovative Approach to Achieving Water Quality Benefits*; Water Environment Research Foundation: Alexandria, Virginia.

Lower Boise Effluent Trading Project Final Report

This report is a comprehensive account of issues dealt with by the Lower Boise phosphorus trading stakeholder group. The report is accessible through http://www.envtn.org/resources.htm.

The Research of Richard T. Woodward

Richard Woodward, of the Department of Agricultural Economics, Texas A&M University (College Station, Texas) has posted a number of papers related to resource economics, including water-quality trading, on his Web site (http://ageco.tamu.edu/faculty/woodward/ paps/).

Useful References

The following references are deemed by the authors to be particularly useful:

Carter, D. L. (2002) Proposed Best Management Practices to be Applied in the Lower Boise River Effluent Trading Demonstration Project. Unpublished report; http://www.envtn.org/docs/carter2002bmps.PDF (accessed April 17, 2004).

Faeth, P. (2000) *Fertile Ground—Nutrient Trading's Potential to Cost-Effectively Improve Water-Quality*; World Resources Institute: Washington, D.C.

Kerr, R. L.; Anderson, S. J.; Jacksch, J. (2000) Crosscutting Analysis of Trading Programs—Case Studies in Air, Water and Wetlands Mitigation Trading Systems. Research paper prepared for the National Academy of Public Administration: Washington, D.C.; http://www.napawash.org/pc_economy_environment/epafile06.pdf.

King, D. M.; Kuch, P. J. (2003) Will Nutrient Credit Trading Ever Work? An Assessment of Supply and Demand Problems and Institutional Obstacles. 33 ELR 10352; *Environ. Law Reporter*, Washington, D.C.

Michigan Department of Environmental Quality (2002) *Rule Part 30: Water Quality Trading* (effective November 22, 2002); Michigan Department of Environmental Quality, Surface Water Quality Division: Lansing, Michigan; http://www.state.mi.us/orr/emi/arcrules.asp?type=Numeric&id=1999&subId=1999%2D036+EQ&subCat=Admincode (accessed April 16, 2004).

Ross and Associates (2000) Lower Boise River Effluent Trading Demonstration Project: Summary of Participant Recommendations for a Trading Framework. Report prepared for the Idaho

Division of Environmental Quality; http://www.deq.state.id.us/water/tmdls/lowerboise_effluent/lowerboiseriver_effluent.htm (accessed April 17, 2004).

U.S. Environmental Protection Agency (2003) Water Quality Trading Policy. Unpublished guidance; http://www.epa.gov/ owow/watershed/trading/finalpolicy2003.html (accessed June 20, 2004).

U.S. Environmental Protection Agency (2003) *Watershed-Based National Pollutant Discharge Elimination System (NPDES) Permitting Implementation Guidance*; EPA-833/B-03-004; http://cfpub.epa.gov/npdes/wqbasedpermitting/wspermitting.cfm(accessed April 8, 2004).

INDEX